Animal Camp

Animal Camp

Lessons in Love and Hope
from Rescued Farm Animals

KATHY STEVENS

Skyhorse Publishing

Skyhorse Publishing books may be purchased in bulk at special discounts for sales promotion, corporate gifts, fund-raising, or educational purposes. Special editions can also be created to specifications. For details, contact the Special Sales Department, Skyhorse Publishing, 555 Eighth Avenue, Suite 903, New York, NY 10018 or info@skyhorsepublishing.com.

www.skyhorsepublishing.com

10 9 8 7 6 5 4 3 2 1

Library of Congress Cataloging-in-Publication Data

Stevens, Kathy, 1958-
Animal camp : my summer with a horse, a pig, a cow, a pigeon, a dog, two cats, and one very patient man / Kathy Stevens.
p. cm.
ISBN 978-1-61608-011-2 (hardcover : alk. paper)
1. Animal rescue--New York (State) 2. Domestic animals--Protection--New York (State) 3. Vacations--New York (State)--High Falls. 4. Stevens, Kathy, 1958---Homes and haunts. 5. Catskill Animal Sanctuary. I. Title.
HV4765.N7S74 2010
636.08'32092--dc22

2010025003

Printed in the United States of America

Contents

About Catskill Animal Sanctuary

C lose to two thousand farm animals, victims of neglect, abandonment, or the food industry, have found safe haven at Catskill Animal Sanctuary since we opened our doors in New York's Hudson Valley in 2003. Through a stringent adoption program, many of our animals find loving permanent homes. Still, at any given time, some 250 animals, ranging in size from two-pound bantam chickens to 2,000-pound draft horses, call CAS home. Over the years, we've watched in delight as newcomers respond to good food, spacious pasture, top-shelf medical care, and love in abundance. For some, the healing is immediate; for others, it takes months, even years, to erase dark memories. "Heal in your own way, at your own pace, on your own terms," is our mantra, and, indeed, of all my roles at CAS, there is no greater joy than participating in the transformation of a broken spirit. The most remarkable among these spirits have become our teachers; the most remarkable lessons have been life-changing.

Through the dedication of hundreds of good people, the derelict property we purchased has been "rescued" as well. A 6,000-square-foot, twenty-stall main barn sits squarely in the center of our eighty acres; atop its green metal roof, two hundred solar panels gather and store all the energy required to power Catskill Animal Sanctuary. The collapsed buildings, piles of old tires and toilets, and discarded automobiles and appliances that we inherited are long gone. In their place are fourteen spacious pastures and twelve roomy paddocks and pens. Six small barns house our pigs, cows, goats, and sheep, while some twenty other outbuildings provide deluxe digs for our horses, donkeys, and smaller animal friends: rabbits and chickens, turkeys, ducks, and geese. Two houses, a staff office building, an organic garden, and a visitors center provide for human needs. Add to the picture the trucks, plows, tractors, and utility vehicles one needs to feed and clean up after the animals, manage pasture, mow grass, maintain a garden, and one gets a sense of CAS: a bustling community composed of animals who routinely surprise and delight us, and humans who feel privileged to serve them.

Each year, I visit dozens of classrooms between Albany and Manhattan, inviting children to consider farm animals in ways they generally don't. If they're young, the questions are simple, but even simple questions prompt important insights. When the children are of middle and high school age, then the presentation can address industrialized cruelty and the suffering it inflicts; it can encourage the tough questions we humans like not to consider. Dozens of school groups also make their way down our hill each spring and summer, and we tailor their experience to each group's age and background as well as to the teachers' curricular goals. These activities, in addition to seasonal weekend tours and presentations at conferences and other events, have comprised our educational program to date.

Now that program grows as well. In the spring of 2010, we launched "Compassionate Cuisine," a series of on- and off-site vegan workshops and courses, dinners, and cooking parties in private homes with Kevin Archer, a chef whose engaging manner and consummate skill inspire confidence in those new to vegetarian cooking—even complete neophytes can't boil the metaphorical vegan egg! We're hiring a director of education to work with schoolchildren both on and off-site, and to develop and oversee Camp Kindness, with both day and "sleepaway" programs for children eager to connect with animals and dig in the dirt. We're building a bunkhouse and converting my house to an education center. By 2012, our educational program will be a year-round operation.

Finally, there's the machine that keeps CAS running: the newsletters and website; the blogs and books and events; the press releases and interviews; and a membership of over 6,000 people from nearly all fifty states and from twenty countries around the globe. There's a whole behind-the-scenes world, masterfully administered for seven years by Julie Barone, whom I refer to as my right hand and my left brain, involving databases and IRS filings and software packages and office equipment and fundraising programs and thousands of details that support everything from our volunteer, animal sponsorship, and membership programs to the proper handling of hundreds of thousands of dollars each year. As any business owner knows, the "back office" is as important, and at least as involved, as the front.

It's a cool Saturday morning as I sit on my deck, laptop in my lap, completing the final proof of this book. Spread out below me in their individual pastures, the cows, horses, and sheep eagerly await breakfast. The roosters crow; a beaver swims diagonally across the pond and waddles onto shore. I once wrote that the joy at Catskill Animal Sanctuary was so palpable that I often felt my heart would explode. That statement is truer today than it has ever been. Seven years after Catskill Animal Sanctuary opened its doors, my vision of a peaceful haven for needy farm animals is a thriving operation, living proof of what can happen when one follows her heart.

Welcome to Catskill Animal Sanctuary

I'll never forget the day that a man and his son, nine or ten years old, arrived a few minutes before Saturday tours began at Catskill Animal Sanctuary. Our vet had just euthanized our beloved rooster Henry as we sat together, Henry in my lap, under the willow tree he loved. A moment later, the two visitors walked into the barn as I was wrapping Henry's body in a towel for burial.

Henry had been a special friend who loved my dog Murphy (he often nestled right into Murphy's side and fell asleep), sat peacefully in the barn aisle observing the daily goings-on, sang to the radio, and loved, more than anything else, to be held. I was going to miss him: we all were. Anyway, the father and son asked what I was doing. Through my tears, I explained that Henry had just been euthanized, that we were getting ready to bury him, and that he'd been a good friend whom I was going to miss. "This is the hard part of our work," I explained to them, then invited them to wait beside the goat field for their tour.

As the two walked out, I heard the son say to his father, "Dad, *it's just a chicken.*"

Seven years into this glorious work of rescuing abused, abandoned, and discarded farm animals, and helping each one recover on his own terms, I love chickens, cows, and pigs more than I could ever have imagined. In fact, something wonderful occurs when one works with animals at a place that's devoted to their happiness. One notices not only behaviors specific to their species, but also characteristics that many people still consider the exclusive domain of humans: the richness and range of their emotional lives, for instance. An angry chicken? Yes. An impatient chicken? Of course. A chicken grieving the loss of a friend? Yep. An affectionate and loving chicken? We've known many. Equally interesting are the distinct differences in individuals of a given species. While there are particular qualities inherent in all chickens or pigs or cows, any ten individual chickens are as unique and individual as any ten humans. The challenge for Catskill Animal Sanctuary is how best to help our community see them for who they are . . . and then face what we are doing to them.

So many wonderful books about farmed animals and farmed animal issues have been published in the last few years that I, for one, can't keep up with all the reading. But despite the harrowing accounts of horrific suffering by the animals we grow to feed humans, despite the well-documented link between a meat- and dairy-based diet and so many of our diseases, and despite the consensus that growing animals to feed humans is the primary cause of global warming, humans are still eating animals to the tune of eighty-five to ninety animals per year per "average" eater. And my theory is that it's in large part because we don't know them. If we did—if we knew them as we know our dogs and cats—then perhaps eating them would feel no less repugnant.

It is for precisely this reason that I wrote another book.

Far and away the best book I know about pigs is *The Good Good Pig,* written by Sy Montgomery, who for fourteen years had a pet pig named Christopher Hogwood. Montgomery shares stories from Christopher's life: of Christopher getting out of his pen, walking into town, and being walked back by an assorted cast of characters; of Christopher's special relationship with neighboring children; of his kindness. I laughed and cried my way through *The Good Good Pig* and strongly suspect that more than a few readers gave up eating bacon and pork chops after putting down this delightful work. Why? Because Montgomery made him real for us.

This, then, is the primary purpose of *Animal Camp:* to reveal these animals for who they are. For if you know them as we do at Catskill Animal Sanctuary, you can't help but love them. And if you love them, some of you will no longer be able to eat them. At least that's both my theory (backed by experience) and my fervent hope. Certainly not everyone who reads this book will change her diet. For most, convenience or habit will trump change. But many humans will change their behavior. I know this from having received hundreds of letters and e-mails from people who've read my first book, toured Catskill Animal Sanctuary, or volunteered with us. In fact, our volunteers are the very best evidence to support my theory. They routinely become vegetarian, and many ultimately become vegan, simply because once you know animals, it becomes increasingly difficult to rationalize away your eating them.

Many pages of my first book were written with Paulie the rooster either sitting in my lap or sleeping next to me with my yellow dog Murphy on his bed right by my feet. Paulie is no longer with us. He died in my arms at the vet's office after stroking my forearm over and over with the side of his tiny head. He knew he was dying; he was telling me he loved me. It's important that I experienced that moment and the thousands of other moments I've had with animals most people only experience as chicken wings, pork chops, and hamburgers. It's my job to speak on behalf of my friends.

The book is organized into two sections. The first, "Animal Camp," describes the summer I moved to my partner David's house, taking with me a cow, a horse, and a pig who were ostracized by their Catskill Animal Sanctuary pasture mates. Having observed herd and flock dynamics for seven years, I wondered whether, and how, the weak ones from their respective herds would bond. Would a lonely horse and a picked-on pig become friends? Would they gain the confidence that they needed to fare better back at home, or would they cower once more as soon as they returned? Whatever happened, my hope was to learn lessons that would inform how we group animals at CAS. Perhaps in the case of outcasts, personality should outweigh species. The second section, "Just Another Day at CAS," begins as "the campers" and I return to Catskill Animal Sanctuary. It's a compilation of essays describing single days or moments at CAS: dramatic rescues and happy recoveries, a long winter day, a child meeting Franklin the pig, what it's like to write while surrounded by our free-range friends. Taken together, they provide a glimpse into a most challenging, unique, most wonderful world, a world in which I feel privileged to participate.

Welcome to that world. May you never be the same.

PART I

animal camp

Get Off the Fence

"Get off the fence," my partner David said to me on a Sunday afternoon in late June. "You've been on the fence for too long."

He was right. Throughout my life, I've made big decisions easily, intuitively—often so easily that I suspect that words like "out of her *mind*" peppered my family's conversation about those decisions. When I was twenty-two, for instance, I crammed my car with things that mattered—music, cartons of books, and a good reading lamp—and headed to Boston from rural Virginia. I knew no one there; I had no job. But Boston called to me and was my home for nearly twenty years. At thirty, I met a man named Jesse Moore, a New York actor, and we moved in together after just ten weeks. And even though my southern family's responses ranged from "I wish his skin were a little lighter" to "Jesus, Kathy, aren't there any nice white boys in Boston?" to "I'd date a black girl . . . I just wouldn't tell nobody," the sky didn't fall, after all, and Jesse and I spent fifteen wonderful years together. By the time I turned down an offer to head a new charter high school opening in Boston in

2000 and decided soon afterward to open a sanctuary for abused farm animals, my family thought I'd gone off the deep end again, but at least knew better than to try to convince me to think about words that started with "s": safety, security, salary, sanity. The only "s" word that's ever held any sway over me is "short," as in life is _____. I've followed my heart my whole life. As far as hearts go, it's a trustworthy one. It has never once failed me.

~~⌇

So why, then, was I struggling with the decision of whether to move to David's house in High Falls for the summer, a mere six animals—three from the CAS crew plus my dog and two cats—in tow? He and I had an easy and wonderful relationship. Its unconventional nature suited us both. We each had our own house, mine on the grounds of Catskill Animal Sanctuary, plopped on top of a hill in the middle of one of the horse fields with a view of the cows, turkeys, goats, rabbits, sheep, and chickens; his on a secluded piece of land just south of the village of High Falls, a quaint community thirty minutes south of CAS. He lived in rural New York for the warm part of the year and in Hawaii for the winter months. We had enough similarities to keep it real and fun, enough differences to keep it interesting, enough space to keep it fresh. And David loved my dog at least as much as he loved me. For that simple fact alone I adored him. Spending the summer at his house should have been a no-brainer. It wasn't.

"This is my place of peace," I said to him one night as we sat on his front porch eating huge bowls of greens picked from his garden. The thought had just come to consciousness. Catskill Animal Sanctuary, the haven for twelve species of farm animals, was my place of joy—the place that fulfilled my deepest need: to give happiness to others.

There, fearful, tentative newcomers learn quickly what we're about and consequently learn just as quickly that it's okay to be what they're about. Chickens fall in love with sheep and pigs raid the kitchen. Yesterday, Zen the goat ripped a chunk off my straw hat. A few minutes later across the big barn aisle, his neighbor, an old horse named Freedom, fell asleep with his head in my arms. Joy is abundant. Love manifests in myriad ways large and small, and we are all the better for it.

Yet there is enormous responsibility that comes with CAS when I'm there. Two hundred fifty animals. One hundred volunteers. Building projects, excavation projects, and fence and road repair. A rabid fox roams through. Beloved animals age; their care becomes more complicated and more costly; their deaths leave us reeling. A nine-page waiting list filled with the pet horses, goats, and pigs whom people in a wretched economy can no longer afford to feed. Cruelty cases and their accompanying stresses. Newsletter deadlines and weekend guests and guest appearances and boards of directors and blogs and summer programming and annual audits and school groups by the dozen and always the need to raise more funds to support our work. Catskill Animal Sanctuary is a joyful place, indeed, but the weight of what I've taken on rarely leaves me.

Except when I'm at David's. There, nestled in a gorgeous meadow bordered by forest, I am surrounded by animals but not responsible for them. There, my breath is deep and slow. That would surely change if our little summer family expanded to include Tucker the cow, Franklin the pig, Hope the horse, plus my own gang of three.

Another reason I was uncharacteristically hesitant had to do with David. David cherishes both his privacy and his solitude, and both were in ample supply at his place. Our mornings begin around 5 AM. The first of us to wake, usually David, makes the coffee and brings it to the lucky lounger. We prop pillows three deep against the headboard and sit up to sip coffee and watch the earth wake. The rest of his day David does what he loves best. I call it "digging in the dirt," for the things that bring him peace are all elemental, all about caring for the Earth and the creatures who call it home. Whether it's tending his garden, sawing down dead cedars to turn into fence posts, planting fruit trees, or stretching hundreds of feet of fencing (at his place or mine) to create a sumptuous pasture for new animals, "digging in the dirt" is work that keeps David's sensitive spirit positive—it is his art and his religion.

David promised me that he wouldn't allow the animals to impact him. "You're going to be the one doing the work, Kath," he mentioned every time we discussed the possibility of Animal Camp. And while that particular fact was true, his sense that his world would not be turned on its head was, well, ridiculous. A happy, formerly single guy was contemplating living full-time with a rabid animal lover and being stepdad to a dog, two cats, a horse, a cow, and a pig . . . for starters. The experience would rock his world; I wasn't sure for the better.

But when the good, good man said one morning, "Come on, I'm waiting for the show to begin!" I had no more excuses. The animals would come.

I called Corinne Weber, our primary hauler, to schedule her for her soonest possible date. It was Saturday, July 11, my birthday.

Choosing Campers

Who are the outcasts—the ones who are always rejected by members of their species? Who would benefit from a quiet summer filled with lots of personal attention? Who is exceedingly kind, might bond with animals of other species, and in the process might offer lessons about how to group animals at Catskill Animal Sanctuary? These were the questions I asked when deciding who would spend the summer in High Falls.

Most of the animals who reside at CAS seem exceedingly happy to have spacious pasture; clean, roomy shelter; friends of their own species; and an abundance of food provided by always kind humans. Add to that frequent grooming and, for the smaller animals, the occasional cuddle in a happy human's arms, and I hope that if they could speak, they would say they were in heaven.

Yet as much love and attention to their individual needs as they get at CAS, some animals either need or wish for more. Hope was such an animal.

She came, this chestnut girl with a big white blaze and four white socks, as one of five mares from a failed Thoroughbred breeding operation. Having been turned out in an enormous pasture to fend for themselves, the horses had never been handled: no halters, no grooming, certainly no veterinary or hoof care. Our wonderful farrier Corey Hedderman called them "semi-feral," and it was certainly an apt description. When they first arrived at Catskill Animal Sanctuary, the mares watched our every move, were terrified of being haltered, and retreated from our advances. One had broken her owner's ribs, kicking out violently as he attempted to cowboy her onto the trailer.

It was Hope who stood out. From her very first moment at CAS, ten-year-old Hope worked hard to get beyond an ingrained fear of humans. In their large pasture, Echo, Dharma, Dakota, and Charmer, her four equine friends, either continued eating their hay or moved slightly away from us when we entered. Hope, though, would raise her head, ears forward, and look up. "Help me trust you," is what that gesture said. "I'm interested, but I'm wary."

And so we would take our time.

"Hey girl," I said to her one sunny morning as I walked into her big field. "Hey, good girl." I stood fifteen feet away; moving closer would be too threatening. Hope lifted her head a foot from the ground, a posture of hopeful subservience, just as she always did. "Humans and I don't have a great track record, but I'd like to try again," is what I read in her body language.

"It's okay, Hope," I whispered as I very slowly turned around and sat cross-legged on the ground, my back to the hesitant horse. This gesture works without fail (and usually quickly) with fearful animals. In an instant it flips the dynamic, making the human the vulnerable one. Suddenly smaller and unable to see what's behind me, I am telling Hope that I will not, in fact *cannot*, hurt her, and just as importantly, that I trust her not to hurt me. I also cannot look her in the eye, something far too threatening for many of our wounded newcomers.

Far more quickly than I expected, I felt warm breath on the back of my neck, then a soft muzzle ruffling the hair on top of my head. Yes, just as I thought. This girl was going to be a gem.

Over the next weeks, as we monitored the horses' diets, removed the burrs from their manes, and helped them overcome their fear of being haltered, the group dynamics in Hope's herd were all too apparent—Hope got picked on by everyone else. She was chased from her food, driven to the outer edges of the herd by the far more dominant Charmer and Echo, and blocked from entering the shelter that was large enough to hold a dozen horses. Several times we shuffled horses, trying to create a herd in which Hope would be accepted. But it never worked. She was always the outcast and had a collection of war wounds—scrapes, scratches, bite marks—to prove it.

Hope would come to Summer Camp.

~⌒

At the opposite end of the farm, Tucker, a young Guernsey steer, was having his own struggles. Tucker had spent the first few months of his life at a petting zoo that, like most of them, purchased baby animals at the beginning of each season to lure in paying customers. At the end of the season, the zoos close down and send the animals to auction and thus, of course, to slaughter, as there is absolutely zero demand for pet cows or other farm animals.

But Tucker was luckier than most. A frequent visitor to the petting zoo had fallen for him, purchasing him to spare his life. A couple months later, Tucker found his way to CAS and was placed in a half-acre paddock right by my house for a three-week quarantine period. I fell in love. Tucker was a 200-pound puppy. He followed me around, ate my hair, licked my face . . . and spent a good portion of every evening standing at the fence, staring at my porch door and mooing with all his might: "I'm lonely! Come play with me!" he bellowed. Often, I would.

Once it was certain that Tucker was healthy, it was time to introduce him to other cows. I knew the main herd would be too much for him—especially young Amos. A four-year-old Longhorn, Amos was a badass. Not only was his attitude intense, so were his horns. They were two feet long and pointed skyward. ("Horny Bastard," I call him, twisting the conventional meaning.) If Amos wanted to hurt the young Tucker, who'd never spent time with cows, he surely could. Tucker would have to join our small special-needs herd of Romeo, a gentle elderly steer; Helen, a blind Hereford; and Dozer, a young goofball and another petting zoo refugee.

From the beginning, it was apparent that Tucker had far more interest in people than in other cows. He was accepted, to a degree, but was clearly the low man on the totem pole. Dozer, smaller than Tucker but both older and more dominant, especially picked on him, driving him away from the herd as it grazed and shoving him away from Helen, the lone female, if he got too near.

"Tucker!" I would call as I approached their large pasture, and the red and white beast would come gamboling at me, then press his head into my belly.

"Love me," he would say. And so I would scratch his cheeks, rub his soft ears, and brush his sleek, growing body.

Okay. Another easy pick: Tucker would come to Summer Camp.

~✦

In my experience, it's tough for the runt of the litter, no matter what the species, to overcome behaviors learned in those few weeks as the smallest of the bunch. My beloved Murphy, for instance, was the runt in a huge litter of puppies. Throughout his life, he's backed down from conflict, avoiding fights at all costs. Franklin was the runt, too. Before he was set aside to starve to death, he'd already developed a defensiveness that he's carried with him into adulthood. Everyone (every pig, I mean) picks on Franklin, because, well, they know they can. Whether it's about who steals whose food, who gets the best spot in the hay at bedtime, who leaves the barn first in the morning, or who gets the biggest mud hole, Franklin is always the loser. Many pigs will stand their ground, lashing out at anyone who tries to take what they either already have or what they want. The result is often a superficial gash in the other pig's thick skin or a torn ear. Franklin has never had a torn ear—he runs away.

Anyone who knows pigs as we at CAS do knows the traits we humans share with our porcine friends. Their emotional range is disarmingly similar to ours. While it is true that most animals feel an exceptionally wide range of emotions, I've never really experienced a cow having a nervous breakdown, a temper tantrum, or a fit of laughter. Cows are generally placid animals. Pigs? Even the most mild-mannered among them grumble and growl during health checks and whine impatiently as they wait for breakfast. The more intense ones would give the finest psychotherapist a run for his money: they are drama queens of the first order.

As for Franklin, he simply wears his heart on his sleeve. An especially sensitive pig, Franklin's greatest emotional need seems to be for acceptance. Yes, he's a needy guy. As our youngest pig, he also still has a fairly intense desire to play, but his insistence is annoying, at best, to our older, far more sedentary pigs. Franklin shares sleeping space with Piggerty, but when they're out in their field during the day, he drives her mad. "Play with me!" he insists by charging at her, poking her sides, doing his funny little pig dance. She whips around and snaps at him—her way of saying: "Could I *be* any clearer here? I'm not interested!" It's deeply sad to me that this exuberant guy has no friends, so I vow to take him to Summer Camp, sensing that a friendship between a horse who lacks confidence and a needy cow and pig just . . . might . . . work.

Moving Day

Take One

I've seen the undercover videos of pork producers loading pigs onto the trucks that will haul them to slaughter, stark examples of the wretchedness of which human beings are capable. The pigs are beaten with metal pipes, bats, clubs, or two-by-fours, or they're shocked into compliance with hand-held electric prods. Terrified for their lives, they rush onto the trucks that carry them to their deaths, then are packed like Vienna sausages for the trip. Pork producers know who they're dealing with. Pigs are famously bright, maddeningly willful, and strong enough to drag a man down the road as if he were a field mouse. Smithfield and other corporate giants have designed every aspect of turning fiercely strong, intelligent animals into bacon and sausage and pork chops to maximize profit by maximizing efficiency and minimizing physical risk to the humans who handle them. Never mind the pigs' *monumental* suffering—the extreme confinement in cold warehouses that is their life prior to slaughter, the wretched journey to their deaths, the things that so frequently go horribly, dreadfully wrong in the slaughterhouse. Never mind all that.

If there is a place on Earth that's the antithesis of agribusiness, it's Catskill Animal Sanctuary. Here, every day is Pigs' Day. Our pigs sleep in heated barns on cold winter nights and wallow in mud on hot summer days. Once in a while, when it's sweltering and we want the pigs to have a treat, we open the gate to the duck pond and watch in delight as 1,000-pound pink packages glide through the water as smoothly as swans. Our pigs are grouped in separate shelters and pastures based solely on friendship, for pigs who don't like each other can do serious damage in the blink of an eye. So, for instance, Roscoe and Reggie are bunkmates, as are Farfi and Jangles. Try putting Piggerty and Farfi, two alpha females, together, and watch the hell that ensues. We clean pigs' ears and eyes, trim hooves and tusks, include flax oil and produce in their diets. When they look a little too pink in mid-summer, we slather them with sunblock. We treat them, I suppose, a little like we'd treat our children. We go to lengths most might consider extreme to ensure not only that their basic needs are met, but that their emotional needs—their varying needs for friendship and connection with pigs and non-pigs—are met as well. A mom wants her child to be well-fed, safe, secure, and happy. Many of us want these things for our companion animals. At Catskill Animal Sanctuary, we want them for all our animal friends: we want them to thrive. This desire is at the heart of all that we do.

～⌒

So we arrive at moving day, which, we know, will be all about the pig. Hope and Tucker should go on without incident. But Franklin?

It's anyone's guess. Instead of electric prods and two-by-fours, we bring, along with plenty of patience, a bowl filled with everything from apples and bananas to thick slices of chocolate cake, none of which will matter a whit if Franklin doesn't want to load.

At 10 AM, Corrine Weber of Equine Emergency Transportation pulls down the driveway and expertly backs her white Ford F-350 and glistening white trailer up to the gentle slope of grass in front of the main barn. (We always get a good chuckle out of her company name, since she's hauled as many cows, pigs, sheep, and goats for us as she has horses.) Tucker the cow and Franklin the pig will ride together in the front compartment; we'll close the solid divider at the halfway point, then load Hope in the back of the trailer.

Sure we will.

Like an old golden retriever, Tucker follows me on first and settles as soon as I place two flakes of hay in front of him. "What a good cow you are," I praise him as I rub his ears and massage his cheeks. Tucker buries his face in his hay. If only it would go this smoothly with Franklin.

Our ammunition—the food—is piled into a feed tub. If all else fails, the heavy artillery—three slices of chocolate cake—is hidden around the corner on the tire fender.

"Come on Franklin, we're going to Summer Camp!" I said, encouraging my friend to follow his nose as I back toward the trailer holding the loot. "Doesn't this smell good?" David, April, Allen, and volunteers Julie Buono and Melissa Bamford are all on hand, and with bodies and voices we steer him toward the trailer . . . for about a nanosecond.

Franklin, you see, wants nothing to do with the big white box or its bovine occupant. He is even less interested than he should be in the feast being dragged on the ground below his snout, because I have stupidly forgotten to tell the staff to withhold his breakfast. Sure, he's a pig, but having just dined on pellets, yams, plums, and berries, he's less interested in our offerings than he needs to be. Time after time, he climbs the ramp, then retreats, climbs, and retreats again.

It's obvious we need the heavy artillery. I dump all three slices of volunteer Phyllis Kaiser's dense vegan chocolate cake into the feed dish.

"Ummpphhhh!" Franklin says, grunting his excitement. "Now we're talking!"

I sit inside the trailer, dragging the lure inches in front of a most enthusiastic snout. When he reaches forward, I back deeper into the trailer. "Ummphh! Ummphhh! Give me the *cake*!" he demands. But I don't.

I am a fool for not thinking this through in greater detail, for not having Franklin contained in a small space prior to Corinne's arrival. While we have never loaded a farm pig before (we've found wonderful homes for many of our potbelly pigs, but not for our 800-plus pound porcine pals), I should have known better. When in the hell has "wishful thinking" ever worked with a pig? In the end, it's clear that our pink pal isn't going anywhere. The longer we try, the more agitated he becomes, and we all know too well that with pigs, the time between agitation and full-out hysteria is about four seconds. We agree to try again as soon as Corinne is available; we agree not to give him breakfast on his travel day.

It takes a few tries with Hope, but she loads in under five minutes. With two-thirds of Summer Camp residents on board, we make our way to High Falls.

Take Two

9:30 AM. Among sanctuaries, the official line is that pigs eat only enough to sustain themselves, and that unlike many other animals, they stop eating when they're full. In my experience, while it might be good PR for pigs, the notion that they're disciplined eaters is utter hogwash. I think it's safe to say that given the chance, most pigs would eat themselves to death. If not to death, then at least to *extreme* discomfort. So today, we hope that Franklin, having missed his 7:30 breakfast, will be more inclined to step onto the trailer than he was last Saturday. In fact, we are banking on that. Electric prods are not an option.

Corinne backs the trailer through Franklin's gate until the rear is a mere five feet from Franklin's door. Using pickup trucks, sheets of plywood, and a metal gate, Alex, Troy, and Cliff, our wonderful summer volunteer, block all possible escape routes so that this time Franklin's only choices are to remain in his cushy barn, deprived of breakfast, or to follow his food dish as it backs magically up the ramp (because I am once again dragging it on the ground just inches in front of his snout) and onto the trailer.

"Kathy, we're ready to load Franklin!" Alex radios me.

I load up his black rubber dish and head down the long barn aisle. *Ahhh*, pig pellets and cantaloupe and kale, with a pint of straw-berries tossed in for good measure. Maybe these will do the trick.

Holding his breakfast, I open Franklin's gate. He hurries out, following the dish that holds the only thing that matters at the moment. "We're going to summer camp, Franklin! We're going to have some fun!" I encourage him with each step. I hold the dish just in front of his face until I arrive at the trailer and place it on the floor.

"Good boy!" I exclaim to my friend as I move backward, on my knees, into the trailer. Franklin looks at me in earnest. "Ummph," he says. And whether it is hunger or fewer people to distract him or fewer options—Franklin can only either follow me or simply stand in place—Round Two is perfect. Just as Tucker had done a few days earlier, Franklin steps tentatively onto the ramp, seems to test its strength, then follows the trail of treats I lay out in front of him until, less than a minute after we've started, Alex instructs, "Come on out quickly, Kathy," Cliff lifts the barriers, and Troy clanks the door shut and pulls the metal lever to lock it into place. Franklin is on his way to summer camp.

A Fine, Fine Spot

David's twenty-acre paradise is nestled a half mile off a rural road and bordered by the 7,000-acre Mohonk Preserve, home to 1,500 plant and animal species, pristine lakes, 2,000 acres of pitch pine, and sheer rock bluffs that provide world-class rock climbing. Though civilization is only two miles away, one hears and sees nothing but nature from anywhere on his land. It would be a fine spot for summer camp.

When one turns into "Forgotten Lands," one drives two hundred yards through woods to a quarter-acre clearing graced by a dozen or so majestic oak and shag-bark hickory trees. Nestled among them is the most beautiful little barn I've ever seen, a barn that reflects David's distinct aesthetic—high ceilings, 10" x 10" posts and beams, open space, huge windows—along with my specifications regarding function. Yes, he built it for us. You gotta love a man who builds you a barn.

"I want the office and the barn to essentially be one big room," I had explained to him during the spring of 2009 as we sat on the porch sketching out our ideas, "but I should be able to close the animals into their own space when I need to."

"Why would you need to?" David asked. Clearly, he hadn't spent enough time with cows and pigs.

"Oh, you know, a cow in an office is a fun *idea*, but things could get messy pretty quickly."

I've always challenged conventional thinking, and this summer would be no exception. Working with animals has confirmed what I've known for some time: that many assumptions humans make about differences between "us" and "them" are based more on prejudice, ignorance, and culturally enforced separation from them than on universal truths about a given species. I wanted to challenge those assumptions. While the purpose of Animal Camp was to build the animals' confidence and discover if (and how) they would bond, I was going to have fun in the process. I would not "train" my horse, cow, and pig friends to enter the office. I certainly *would* open the door and see if anyone *chose* to enter, and I certainly would learn something from those decisions.

Accordingly, the structure, made of pine, red cedar, and hemlock, is eighteen feet deep by twenty feet wide: one big room is both barn and office. The floor is made of thick hemlock planks, a hatch cut into the floor opens to a full basement, and an airy loft is above the main floor. An "open-air" stall occupies the front left corner of the building and opens fully to a paddock; a gate can be closed to keep animals inside the stall. The other three sides of the stall are low walls, so that any visiting animal can stretch his head outward on the left into the fresh air, on the right into the office, or in the back to the feed bins just beyond his reach. It's a wonderful design with ingenious simplicity. The animals will be able to hang out on one side of the barn, I at my desk on the other, divided from them only by a four-foot wall. While I'm at the computer, I'll be able to touch a warm muzzle or a cool snout. By design, there is plenty of room for mischief and mayhem.

The barn paddock is about a quarter acre and is connected via a gate to another of the same size. Three run-ins (three-sided shelters with open fronts that allow animals to come and go at will) are enclosed within this paddock. Both paddocks are shaded by towering trees, and both have gates at their far sides that open to a large, sunny pasture that stretches all the way back to David's house, located a thousand feet behind the barn. If necessary, say at feeding time, this "two paddocks and a pasture" design will enable each animal to have his or her own enclosed space.

From the barn, one continues down the driveway to David's home, built with wood milled on the property and designed with the same aesthetic that informed the barn: post-and-beam construction, high ceilings, huge windows. The house sits roughly in the middle of the property in a large, south-facing meadow. Just in front of it, two small ponds attract turtles and frogs and birds— mosquito-loving phoebes and flycatchers and red-winged black-birds. Bluebirds make their nests in the tall grass of the meadow, goldfinches feed on the purple catmint that grows near the pond, and great blue herons, having fed from the stream below, soar over-head. Deer, foxes, and bears wander through.

Five hundred feet from the house is a wall of mature pine, hemlock, and cedar trees. David owns the first hundred feet or so of these deep woods; step beyond that and one has stepped onto Mohonk Preserve.

On the left edge of the meadow, about fifty feet in front of the porch, is the first of two small ponds, some sixty feet long and fifty feet across. This pond is flanked on the left by a graduating climb of quartz/granite conglomerate boulders ranging from those the size of washing machines to big ones the size of school buses. A large boulder juts out over the water about fifteen feet above the pond's surface. This one is David's diving board. Every hot summer day, he scales the smaller rocks that rim the pond, steps onto the flat surface of the diving rock, and dives into the cool water below. This pond is also Murphy's most frequented swimming hole. (His Mom's, too.)

Beyond this pond is a kidney-shaped one approximately half the size, and behind that, a full acre of berries that we share each July with resident and visiting wildlife. On the right edge of the meadow is the southern end of the animal pasture. My hope is that while David, Murphy, and I are relaxing on the porch on warm evenings, our three friends will venture from the barn to join us. I don't tell David about my other hope: that Franklin will join us for a swim.

Hi, Friends!

Hope and Tucker were at the gate when we arrived, and as Corrine backed the trailer to their paddock, they were all ears and attention. Unless a cow is exploding with glee (in which case they do a little dancy-dance bucking and kicking and twirling around their pasture—how delightful it is to watch!), it's generally easier to read excitement in a horse or a pig than it is in a cow. But today, Tucker's insistent *moo-oooo!* told me that his curiosity was nearly killing him. After all, from his point of view, he and Hope had arrived at David's in this same big box a few days earlier: who was in there now?

"Franklin!" David and I call to our porcine pal. "Welcome to Summer Camp!" I did it to comfort Franklin, for surely this sensitive animal was filled with uncertainty and trepidation. David called out to him in sheer excitement, for David has adored Franklin since Franklin's first night with us, when as a four-pound pig he fell asleep between us, snuggled into the blankets of David's big bed.

For the safety of all, David moved horse and cow into the bordering paddock that contained their shelters. They'd have to watch the show from a few yards away. Tucker was none too pleased with this arrangement.

I swung open the trailer's heavy metal gate. "Hello, my love," I whispered to my friend, who turned around from his forward-facing position and ambled toward me. "You made it! Come on out and look around!" He was shaking.

The trailer that Corinne used today is called a "step-up" trailer. There's no ramp to walk up; instead, both human and animal must take one very large step, probably a good 14 inches, to get in. At CAS we pull homemade ramps up to the trailer to ease an animal's entry; until now I hadn't even thought about how difficult it would be for a pig, with short legs that barely bend, to take a huge step down from the trailer to solid ground.

"Oh, wow," I said, looking at the big step, then at Franklin's little legs. "This is going to be hard for him." Ever resourceful, David ran to the lumber pile for a makeshift ramp. Franklin, meanwhile, had frozen in place. It's what pigs do when they're trying to size up a situation and make a decision. By the time David had retrieved a half-dozen boards with which to fashion a step, Franklin had already solved the problem. He simply turned sideways, his body parallel to the opening, and lay upright as closely as he could to the edge of the trailer, his front legs outstretched. Then, with the bulk of his weight resting on the trailer floor, he scooted forward with great effort and placed his left foreleg on the ground. He scooted again and placed his right foreleg.

"You're a smart boy, Franklin," I praised him. Franklin rotated his body until his butt sat on the edge of the trailer, then stepped down with one rear leg, then two. He walked several steps, then lifted his pink snout to the air to take in his new world.

"Look Hope, look Tucker," I encouraged the red and white horse and the red and white cow. "This is your new friend." Hope was a nanosecond from fleeing for her life, but, summoning her courage, stood in place, her eyes growing larger by the second.

"Moo-oooo!" Tucker called.

Franklin, though, was more interested in his home than his housemates. After all, this pig had been picked on his whole life. As the runt of his large litter, Franklin was battered about in the frantic search for one of mom's teats or the favored sleep spot. When his savior (a slim woman with dark hair whose name I don't recall and wish I did) found him, he'd already been separated from his family and tossed aside to starve to death, a common practice on farms that grow pigs to turn them into bacon, pork chops, and sausage. When the tiny piglet arrived at CAS, his instinct to retreat in the face of the simplest threat was already deeply ingrained. At the slightest hint of conflict, Franklin scurried away. Today, his snout reaching no higher than the elbows of the red horse who towered above him on the other side of the fence, Franklin was keeping his distance.

A whole new world was laid out before him. Water and ferns filled a low spot thirty feet from the barn entrance, and the moment he saw it, Franklin was a pink tank moving toward it. He rushed into the water and shoved his snout to the ground. Up came a face covered in cool brown mud. Again, this time deeper, and again. Franklin's snout was a shovel, and each time he lifted his face, he loosened more heavy wet earth until the space was precisely the way he wanted it. He dropped his body down into brown, goopy bliss. For the first time in his life, Franklin had his own private mud hole. Behind the safety of her fence, Hope snorted.

"Pig dipped in chocolate," David laughed as a few moments later Franklin stood to continue exploring.

"Yep." Franklin's left side was a cool pink, his right was the color of milk chocolate.

"Ummmph," he snorted a greeting as he walked past us.

"Hey, happy pig!" I responded.

For the next few minutes, Franklin explored every inch of his paddock. He smelled the pine trees, rooted at the edge of the fence, blew bubbles in his water trough, and discovered a new delicacy: hickory nuts. For a good ten minutes he searched for them, inching forward, snout to the ground. When he found one, he'd gobble it up, taking time to crack the shell and spit it out before chomping on the tasty nut. He moved around this new home, taking it all in, *ummpphhing* his approval with every few steps. "What do you think, Franklin?" we asked him.

"Ummmpph."

Soon, though, his misgivings notwithstanding, Franklin was curious to meet the large beings locked behind the gate. I stood up from my perch on the barn step and walked over to encourage the introduction.

Red horse head reached over the fence as pink tank approached. Hope's neck was arched, her ears were forward. She was anxious, as she was with us humans during her first days at CAS, but she was eager to connect, the desire to say hello transcending species. (For Hope, this seems universally true. If you are alive, this horse wants to know you. At least that's my hunch. And yet that desire is rarely satisfied. For whatever reason, whether genetics or history or both, fear always supercedes desire. Hope is timid. The more aggressive being—horse, human, pig, cow—controls the moment. Hope, like Franklin, retreats.)

And then there was Pink Tank. My dear Franklin, who had to overcome three years of being jostled and jolted by porcine pasture mates. Who had to overcome being bitten by Maxx the horse when he strolled through the barn. Did Franklin have the capacity to get out of his head for a time and simply let the moment happen?

I am delighted to report that yes, indeed, he did. Muddy pink snout lifted, soft black muzzle lowered, and in a moment so pure and innocent that it is forever etched in my memory, two creatures abused by those of their own species said hello. They simply breathed, taking each other in. Nothing more than that; but for these two, it was an auspicious beginning.

Lola

I am filling one of two large water troughs when our friends Mary and Lola pull in. It's been two days since Franklin's arrival. Lola, a scrumptious three-year-old, has been to CAS several times. Now, however, Franklin is practically her neighbor; Lola and her parents, Aaron and Mary, live just a couple miles up Mohonk Road.

"Lola!" I shout as I walk toward their car. It has been a few weeks since we all had dinner together. Like many young children, she had taken a few minutes that evening to get beyond her initial shyness, but by the end of the night, Lola had very carefully painted David's fingernails a lovely sparkly pink *and* invited him to a party. I was green with envy. David had driven Lola to that party (in the next room) about ten times, and each time they returned (after about four seconds), the pair brought back a friend whom they'd met at the party (a random item from the room). Eventually, Lola tired of the game; we said goodnight to our friends and fell asleep imagining Lola's first visit to Summer Camp.

Here she was, feeling shy again, burying her face in Mary's neck.

"Lola, would you like to come see *Fanklin?*" I ask, pronouncing his name without the *r*, the way Lola does.

(Franklin loves children; Lola remembers him from CAS. She and Mary had actually given me a beautiful portrait of Franklin for my birthday that Mary, a talented artist, had painted. I also got birthday bubbles from Lola—*exactly* what I wanted.)

Anyway, not a peep from the child.

As Mary and I chat, though, Lola soon signals to Mama that she wants to see *all* the animals. The three of us walk into the wonderful open air barn, where Hope is resting in the only stall, separated from us by a four-foot fence. I climb over it, inviting Lola to come, but Hope's mass is too much for her. She stays in Mary's arms, and the good horse knows what she needs to do. She presses into the fence, stretches out her long neck and gently touches Lola's calf. Lola stiffens.

"It's okay, Lola," Mary encourages her. "She's saying hello."

Slowly, gently, Hope reaches up and nuzzles Lola's cheek with her warm muzzle, signaling that she's really just love in a thousand-pound package. It works. Actually, it takes but a moment before Lola is on the other side of the fence, brush in hand, grooming the sleek red body of the animal who towers above her.

Tucker, known by Lola as "Tucka," is hanging out just outside the barn; his legs curled under him as he chews his cud. It is striking that despite a field filled with lush grass, Tucker almost always chooses to be near me. If I'm in the barn, it's where he wants to be. His preference, in fact, is to be in the office, right next to me, but his office skills leave a little to be desired. I have to reserve his visits to days when I have time to remove anything that he might wish to lick, chew, or knock to the ground with a single swoop of his big head.

"Hi, Tucka," Lola says to her red and white friend, and we walk over with some bananas that I've grabbed from the fridge.

Meanwhile, Franklin is taking in the action from atop his bed of straw in a nearby shelter. As we move toward the water trough to finish filling it, he barrels over, steps gingerly into the trough, and sits down, sloshing most of the water out of the trough and instantly muddying what remains.

"Franklin!" I say to the imp, who is so clearly showing off. "This is not your bathtub!"

When Lola laughs, I turn to her. Forcing earnestness onto my face, I say, "Lola, this is not good. This is water for Franklin and his friends to drink. I think Franklin needs to go to 'time out.'"

In an instant, Lola's delight changes to anxiety. Mary shakes her head "No," and explains that "time out" terrifies her.

"Oh, Lola," I say by way of apology. "I'll bet you have a different idea. What do you think we should do?"

Lola and I have a lot in common, one of those things being the volume of our voices. Truly, the child could wake the dead. Her suggestion is instantaneous and resolute: "We should say, 'Don't get in the bathtub, Fanklin! *Don't get in the bathtub, Fanklin!*'"

With that, Franklin steps out, turns around, roots under the trough, and very deliberately tips it over. We laugh together as I walk them back to their car.

～◯

It's not that Lola's speaking voice is loud. Actually, her speaking voice is that of a normal three-year-old. But when something feels exceptionally important to her, bellowing it is her way of conveying its weight. For the past few weeks, her pronouncements to David and me all have to do with diet.

For instance, when the phone rang one morning, we heard Mary say, "Lola has something important to tell you," before she handed the phone to her daughter.

"I hear you have some important news!" I said. "What is it?"

"*I don't want to eat meat any mo-wa!*" came the response.

After I picked myself up off the floor, I said, "Lola! That's wonderful! Why did you make that decision?"

"*Because I love animals!*"

A few weeks later, ten of us were celebrating David's birthday on his wonderful front porch. I turned to Lola, sitting between her parents.

"Hey, Lola . . . are you still a vegetarian?" I asked.

"No."

"Oh," I said. "Why did you change your mind?"

"*Because I like chicken!*"

Having said goodbye to Franklin, Lola is climbing into the Subaru's back seat when Mary says, "Lola, want to tell Kathy your big news?"

"*I'm vegetarian!*" Lola hollers from the back.

"I am, too!" I shout back from just outside the door.

"*I love you!*" Lola shouts.

"I love you, too!" I yell back.

And then they are gone.

While in my perfect world both parents, and hence the child, would be vegan, I love that Mary and Aaron are encouraging Lola to make her own choice, and that they're honoring the struggles between a three-year-old's heart and her taste buds. I love that this tiny sprite is wrestling with such a huge question—*Do I really want to eat them?*—a question that too few of us ever consider. Lola lives in a home where love abounds. She has parents whose every action encourages her to discover and refine her own values. I am so happy that Fanklin and Tucka are just down the road to offer their support.

Glory Days

It's seven o'clock on a beautiful summer morning when Murphy and I enter the barn. "I'll be back, pooch," I say to him. Murphy settles onto his bed, grabs a towel, and begins to gnaw.

I walk to the big pasture. No hint of anyone, which is understandable: we're half an hour early. Normally, the threesome is hovering around the office (which doubles as their feed room) when we arrive at seven-thirty. When we're early, like we are today, I have to call them in.

"Hope! Tucker! Franklin! It's breakfast time!" I holler as loudly as I can. "Come on, animals! *It's breakfast time!*"

Hope trots through the narrow, wooded lane that connects the two ends of the animals' large pasture. She's always first, of course: she's a Thoroughbred after all. (Franklin may well be more food-driven, but his twelve-inch legs just don't allow him to cover ground as quickly as a racehorse can. Damn biology.)

As soon as she enters the field, Hope breaks into a canter, then into a full-out gallop. She's already covered the three hundred feet between the lane and where I stand and has zoomed past me and planted herself in the open stall when I spot the pig and cow pair trotting through the lane.

For a moment they're side by side. Tucker is all goofy-cow, and even though he's some three hundred feet from me, I laugh. When Tucker moves at any speed faster than a walk, in fact, I always laugh. It's not that cows necessarily look funny when they're moving quickly—it's that cows moving quickly when they're excited are, well, hilarious. They don't simply run. They run and buck and toss their heads and kick their rear legs sideways and try their damndest to do it all simultaneously. So anyway, here comes Tucker, hauling ass. Beside him, at least for a hundred feet before Tucker can outrun him, is Franklin, running as fast as his little stump legs will allow. I stand there, urging them on. It's a delightful way to start the day.

As I open the first feed bin, the phone rings. It's Abbie, our animal care director.

"Heather is coming this morning to X-ray Mango's hooves. Do you want her to look at anyone else?" she asks. Mango was rescued from a Thoroughbred breeder who had no hay to feed over sixty horses on his property. She is barely able to move from the pain; we are worried about the diagnosis.

"When's the last time she looked at Cas?" Cas is another Thoroughbred. He hasn't gained a pound since his rescue over a year ago. Recently, his behavior suggests he's in some degree of pain. We've no earthly idea why. We've floated his teeth, tried every dietary weight-gain regimen we know, screened and tested for every disease or condition known to veterinary medicine. We've tried acupuncture and acupressure and both homeopathic and conventional treatment for ulcers. Three conventional vets and one alternative vet later, Cas is *losing* weight. He shifts from side to side, he yawns compulsively as if to release stress, and his coat is a mess. "Failure to thrive," is what medicine might call this, since no one can determine what's wrong.

"It's been a month," Abbie says, and we agree that she should look him again today. The decision is a last stab at hope.

April calls in at 7:38 AM. She needs to know the status of our hay delivery. It's been a wet summer, and farmers are suspecting that they'll only be able to make hay once, instead of the normal two cuttings per season. If the supply of hay is half of what it usually is, everyone who needs it will be in trouble. CAS will be in big trouble: in the winter, we use a tractor-trailer load every two weeks.

"Can we order some stone?" Troy asks when he calls at 7:50. He tells me the electrician is coming this morning, and that the carpenters are finished replacing rotten soffits and fascia around the perimeter of the barn.

I say "yes" to the stone, "great" to the construction, then turn to the fridge and pull out the five-gallon bucket marked FRANK-LIN'S FOOD, and finish slicing the morning's offerings: cantaloupe, banana, summer squash, chard, strawberries. I'm late getting started, and the troops are registering their complaints.

Hope leans over the three-foot rail, stretching her neck and head as far as they'll go and when they'll stretch no further, flapping her muzzle for a treat. She sees the kale, summer squash, peaches, and watermelon that I'm chopping for Franklin, and she wants, well, all of it. I've never known a horse as eager to try new tastes as Hope. She's a true vegetarian gourmand. I slip her half a peach.

Just behind her, Franklin is *ummphing* every few seconds, and the longer I take, the louder he becomes, until he is no longer *ummphing* but squealing insistently, his brown eyes wide with anticipation. Tucker, meanwhile, is doing what cows do better than any other species: he's staring at me. His head, too, hangs into the room. He's hoping for a treat, for in his six weeks at Summer Camp, he's discovered, for starters, kale, Swiss chard, plums, mangoes, bananas, strawberries, and whole wheat organic tortillas. Tucker has perfected the art of respectful begging. He knows that all he needs to do is stand at the gate, just six feet from the magic white box with the low hum out of which emerges the food of the gods, and stare at me with huge brown eyes and all the earnestness and hope a cow can muster until I am no longer able to resist. We both know the game.

"Okay, Tucker, yours is coming," I say, and I take him a bunch of Swiss chard, offering it one huge stem at a time. He is delirious.

Alex the homing pigeon is pacing between Tucker's legs, knowing it will be her turn soon. She appeared nearly three weeks ago. In my experience, that's *way* longer than homing pigeons typically spend at a rest stop en route to their final destination. I suspect she's quit her job and has simply neglected to tell the boss.

"Alex," I say to her, "good morning, beautiful bird!" With a quick tilt of her tiny head and an upward glance at me, she makes clear that she knows I'm talking to her. "Breakfast is coming, tiny girl." For the last three days, Alex not only no longer flies a few feet away when I come near; she's begun to walk with me when I go outside. *David does not know this.* He would rightly insist that there was no homing pigeon on the list of camp applicants. He would rightly insist that pigeons are messy, and I would agree with him . . . then secretly hope that today would be the day that Alex jumped on Tucker's back to go for a ride.

In the midst of the activity, Murphy is lying on his plush bed, strategically placed for optimum viewing. Though at thirteen his vision isn't what it once was, he still sees plenty well to watch my every move. I drop a slice of melon (accidentally, of course) and my old pup leaps, grabs it, and gleefully returns to his post. "Good job, dog," I praise him. "Good helping Mama." And then I accidentally drop a hunk of banana, a chunk of squash. There's no better meal-prep assistant than the Great Dog Murphy.

I place Hope's dish in the barn stall, and Franklin's squeal rises to a deafening pitch. He knows the routine: Hope first, Tucker second, Franklin third. The squeal is how he manages the wait. If he couldn't squeal, he'd probably need to plow down a tree.

I reach over the low office wall and carefully toss Tucker's dish to the ground. He dives in.

"Pooch, I'll be right back," I say to the mutt as I grab Franklin's dish. "Guard the barn." Murphy balls his towel between his front legs and goes at it, gnawing himself into oblivion.

Franklin follows breakfast to the paddock that houses the run-ins. Here, he can relax as he eats. About a week ago, Tucker figured out that Franklin is a 700-pound pushover, and as soon as he finished breakfast, he would charge over to Franklin's dish and dive in. Franklin would stand his ground for a moment, but would eventually retreat. So now Franklin eats alone, and like Murphy, he literally licks his bowl clean.

I grab pitchfork, shovel, and rake, and while my friends are feasting, I pick up poop deposited the night before in the largest of the shelters and toss it through the window into a pile that David will retrieve with the front-end loader at the end of the day. There is horse manure, cow manure, and in the front corner of the shelter, pig manure. I smile. David designed Summer Camp so that each of the three large animals could have its own sleeping space, but last night, during a violent thunderstorm, they chose to share a spot.

With Murphy at my side, I walk the yard, cleaning cow pies plopped helter-skelter, picking up branches downed by the storm, and noting the proximity of a certain one-pound pigeon. Yes, Alex is following us. She's aware of Murphy, of course, and keeps a safe distance, but she somehow seems to know that we're all on the same team.

"I won't tell David, girl . . . I promise," I say to our new friend.

∼◦

Both immediately before and immediately after breakfast and dinner, one of two things happens: either the three friends graze together in the large shared pasture, or, if I'm working in the office, they hang around, often wanting to be as close to me as possible. Frequently, two of them will share the stall just next to where I'm working, and the third one, usually Tucker, will stand just outside the office gate, hoping to be let in. I'd let all of them in far more often than I do if there weren't an obvious downside to having a horse, a cow, and a pig as office assistants. Hope and Tucker mess with everything: the grooming box, my stack of papers, the halters and lead ropes, the phone, my laptop; Franklin, though he's generally well-behaved, is still a pig, and could easily flip the table that serves as a desk in an effort to get to the feed bins just beyond his reach. When the animals are in the office, let's face it: I don't get too much work done. Mostly, they have fun while I put the computer out of reach, encourage their exploration, and acknowledge that it's a good thing that I don't have children. I'm sure they'd all be criminals.

It's a muggy day, and Franklin soon heads for his mud hole. I walk over with the hose to fill it with more cool water. First, though, I spray his pink body, and Franklin grunts his pleasure. "You're welcome, pig!" I say to him and watch as he lowers himself into the cool earth.

Both Hope and Tucker are clammy—a perfect excuse for bath time. Six weeks ago, the hose frightened Hope; Tucker would hightail it away from me as soon as I turned the water on. But that was six weeks ago. Now, bath time is a big event. I gather shampoo and grooming tools and place a blue halter over Hope's red head. She doesn't want to move. She's in the stall right next to my work space; it's her favorite. With a little encouragement, though, she steps outside. I pick up the hose, adjust the pressure, and begin at her hooves. Quickly, I work my way up the sleek body: leg, shoulder, neck, back, and rump. I spray with the left hand, massage with the right to loosen the sweat. I don't need to hold her for her to be still. Hope loves this. I move to her head; this has become her favorite part of our frequent ritual.

"Hey, good girl," I encourage her as I adjust the pressure again to a fine spray. I hold the hose just a few feet in front of her and direct the mist to her forehead. She closes her eyes and stands motionless, enjoying the experience, so I adjust the pressure upward and massage her ears, her forehead, and her eyes as the water runs down her face.

A few minutes later, Hope has been scrubbed, soaped and rinsed. I praise her, remove her halter, offer a banana—her favorite, peel and all—and watch the inevitable as she trots to the pasture, rolls on the ground, and is instantly filthy.

Tucker has been standing nearby the entire time and allows me to take the same blue halter and place it over his white and red head. This simple act is a victory in and of itself, but it's nothing like giving a cow a bath and watching him love every moment, particularly when just three weeks ago he ran for his life whenever I picked up the hose. I spray a hoof as he chews on my shirt sleeve, and as I move the spray up his legs to his neck and shoulders, he stretches my shirt until it could fit a linebacker. "Stop, Tucker!" I say, but I'm laughing the entire time, so it's not like he's actually paying attention.

~

Later that night, David and I are sitting on the porch, watching the sunset and sipping tea. Murphy is in the grass just a step below us, sniffing the air for deer. He knows they settle all around us for the night: on the cliff to the left of the house, in the berry thicket beyond the two ponds, and just inside the tree line of the woods. He used to charge after them, interrupting their nighttime bedding down; now, he's generally content to lie in the grass, taking in the sights and smells of a warm summer evening.

A sound to our right is far too loud to come from a startled deer. We look over at the pasture, and here come the three muske-teers: Hope and Tucker are running; behind them by fifty feet, Franklin is doing his best to keep up. They come right to the gate just fifty feet from David's porch, form a line, and look over at us. David retrieves some bananas from a bowl in the kitchen, and the three of us walk down to greet the three of them. We are unlikely friends, but we are friends nonetheless.

Reflections

The purpose of Summer Camp was to see how three "victims" constantly harassed by their own species would coexist, and how that experience would change them. It was to figure out whether it is sometimes preferable to group animals by *personality* rather than by species and to take back to CAS whatever lessons I learned that might impact how our animals live. Selfishly, it was to gain some private, one-on-one time with animals that I craved, time that was rare on Sanctuary grounds, where I am pulled in a million directions. After eight weeks of Summer Camp, we close up the barn and return to CAS. I am delighted to share what I have learned.

Within a couple days of Franklin's arrival at camp, Hope took on the alpha role, and while she used her newly gained status at feeding time, she also used it anytime the gangly, rollicking Tucker approached Franklin a little too assertively. "Play with me!" Tucker insisted as he charged at Franklin, head down, legs whirling through space. Did Hope sense Franklin's panic? Did she empathize with him? Did she know from her own experience as the underling what it felt like to be charged? All I know is what I witnessed. Time after time, Hope simply placed her body sideways in front of Franklin, physically preventing Tucker from intimidating him. Tucker quickly got the message. It seems the maternal role suits Hope well.

Hope will return to CAS a more confident and trusting horse. The uncertainty in her eyes when one approaches her is gone, as is her dislike of having a halter placed on her head. I can pick up her hooves to clean them and lean into her as she naps in the shelter. I can bathe and groom her and watch her body relax rather than stiffen. Yet far more than from the one-on-one time with me, Hope has benefited from being the undisputed leader of her tiny pack and sharing her days with two pals who simply enjoyed her company without an agenda. I wonder if she will carry her newfound confidence with her.

Had the sensitive horse not been here to figuratively hold Franklin's hand, I'm not sure Franklin could have broken through his fear of Tucker. Prior to this summer, Franklin never had a protector. Quite the opposite, in fact. Every larger and more aggressive pig would come after him, and Franklin, still the runt despite his adult girth, would retreat. Now, though, Franklin knows in his bones that neither Tucker nor Hope will hurt him. How do I know this? I know this when I see the three of them grazing in a tight cluster. I know it when Tucker comes barreling right at his porcine pal, and Franklin stands his ground. I know it when the three of them are jammed into the tiny run-in David designed for a single animal, Franklin lying in a mound of hay; Hope and Tucker hovering over him. For the beleaguered Franklin, this level of trust and confidence approaches miraculous. It remains to be seen whether this newfound confidence will serve him in a group of pigs.

And Tucker? Hope has taught him the concept of limits. Hope has taught him to slow down and take a breath. Tucker understands that he can't play with Franklin in the way that young steers like to play. He's tried. These days, he approaches Franklin slowly, stretching out his neck until white cow nose and pink pig snout meet. They stand that way for a few seconds, saying hello, breathing in each other's innocence, Tucker so clearly saying with this gesture of kindness, "I'm your friend, pig. You're safe with me." Clearly, Tucker has also benefited from the unique nature of his summer pack. At Animal Camp, there's no vying for a place in the group like there is at CAS. The group just *is*—an affable, uncomplicated unit.

Tucker and Hope nap near Franklin as he squishes even more deeply into his mud hole. They spend a good deal of their summer days napping, in fact, and while they sometimes do retreat to their individual shelters, they just as frequently cram into the one designed for a single pig. All day long, whether grazing, resting, or simply "being," these three keep each other company, respect each other's limits, and prefer closeness to separation. In terms of the animals' happiness, Animal Camp has been a resounding success. So, too, regarding the larger purpose of the experience. If one can generalize from a single experiment—if Tucker, Hope, and Franklin have proven that in some cases, personality, and not species, should determine who lives with whom—then we've got some thinking, and perhaps some animal rearranging, to do.

As for me?

As I start up my computer after bath time, ace office assistant Tucker wanders in to begin his morning duties: licking my laptop, knocking stacks of paper to the ground, rubbing his chin against the hard-bristle brush in the tack box. Alex, perched on the windowsill no more than a foot from Tucker, does not fly away. A breakthrough.

Thirty-eight new e-mails: a dinner in the city with a prospective new board member, a lunch meeting with our *pro bono* financial advisor, comments from readers of my first book, advice sought by a woman who wants to start a sanctuary, confirmation of a speaking date, a message from someone wondering if we'd be interested in a "barn dog," and so on. The one that stands out is a plea from a sanctuary in foreclosure, wondering if we can take six horses, a donkey, a calf, four pigs, twenty ducks, and "about fifty" chickens. I leave a message on her voicemail offering to take the calf and chickens and help her find homes for the rest.

By the time I finish with the morning's e-mail, it is nearly 11. The phone has rung another half-dozen times, and it's time to leave for my meeting with Troy, whose growing list of buildings and grounds projects needs our mutual attention. At 2 PM, I have a meeting with Bill Spearman, president of our credit union. At 4 PM, it will be time to feed and clean up the Summer Camp barn, and between now and then, thirty more e-mails will have made their way through cyberspace and be waiting for my attention. My book, this book, is due in three months. If I am going to write for a significant chunk of time today, it will be at 7 o'clock tonight, when what I'd really like to be doing is getting into a steaming tub, then crawling into bed with the man, the mutt, and a movie. And you know what I wanted to do today? I wanted to play with the animals; that's about all.

So, regarding the "as for me" question, Animal Camp has offered several important insights:

1. I'm really, *really* good with animals. I know this about myself, but as CAS grows and my responsibilities keep me away from the barn, it's been wonderful to have this reminder.
2. Like some people crave chocolate, I crave meaningful time with critters. Something elemental in me needs it.
3. I absolutely love my life, but I sure would like to simplify. As for David, he survived.

Returning Home

As if she sensed our departure, Alex the pigeon left yesterday. I am deeply concerned for her. Did she get taken by a hawk? Did she return to her city rooftop home? Is she a racing pigeon? Selfishly, I am saddened: I will miss the little sprite and the lessons she offered.

Meanwhile, both Hope and Tucker traveled quietly back to Catskill Animal Sanctuary. Tucker once again loaded like an angel; Hope required a little more patience but eventually took two quick steps up the ramp onto Corinne Weber's trailer and settled quickly. Franklin will stay behind; we'll come for him as soon as his pals are safely settled at CAS.

We're placing both horse and cow in the large turn-out pasture just behind the barn—the field where the special-needs horses, those with hoof or eye or age-induced issues, spend their days grazing peacefully together. At dinnertime, we retrieve them—Noah, Star, Freedom, Abbie, Lexie—and they return to their individual barn stalls for the night. It's a kind, benign group.

What a delight it was over the summer to watch Hope's tentativeness and insecurity disappear as she lived with a young steer and an affable pig. Will this group of special-needs senior horses be equally accepting? And if this happens, will she help Tucker integrate with the horses? Other than alone, we've nowhere else for him to be—he'll be ostracized by the larger or more aggressive cows just as he was prior to his vacation.

I back Hope halfway down the ramp and hand her lead rope to Allen. I take the gangly cow, who leaps the ditch and half drags me to the gate. Thankfully, the other horses are at the far end of the field, oblivious to the goings-on. "You're a beautiful girl," Allen says before releasing the mare; I hug my cow before I let go. "I love you, Tucker," I say to him as Abbie and volunteer Julie Buono join us in the field.

Hope eyes the horses two hundred feet below her. She arches her neck, lifts her tail, and trots expectantly in their direction. At once, they move forward to meet her. First, Noah. The others follow closely behind, and in an instant we're watching an equine dance improvisation: the nose-to-nose greetings, the circling and snorting, the charging away and coming back. I am struck that there's not a single kick, not a single squeal; instead, just the raw beauty of a group of kind horses connecting with the newcomer.

Tucker, who has wisely moved away from the commotion to the fence, is beside himself with excitement, and when the herd takes flight as a single group, Tucker barrels forward to join them, head low and back legs kicking out in goofy, unbridled cow joy. But then I watch the worst possible thing: Hope turns on him.

She charges menacingly, ears back and teeth bared. She means business and Tucker knows it. He backs off.

The horses settle within a few minutes, seven heads down to graze. Tucker once again draws near, but then three horses, one of whom is Hope, come at him, and he is suddenly running for his life, right toward the humans. Yes, Tucker is racing toward us with the three horses right behind him.

Humans rush forward with menacing voices and hands in the air to disperse them. "Tucker, I'm so sorry!" I call as the young cow careens forward, stops dead, then attempts to hide his 600-pound body behind my 120-pound one. Satisfied, the horses saunter off, leaving the frightened, dejected young steer, who lowers his head, pressing it into my calves, asking for a good rub-rub-rubbing of his cheeks, his forehead, his ears . . . the kind he got every day at Summer Camp.

When I enter the field a half-hour later to check on my friends, I have a currycomb in my left hand, a brush in my right. Hope is in the middle of the herd, red head deep in late summer grass. Tucker is grazing by the pond, a good three hundred feet away from the pack of horses. I call to him. He looks up and walks toward me, uttering a low, baleful *moo-oooo*.

"My little man," I say to him as he lowers his head for another rub. "This won't do, will it?"

"Moo," he responds. "Moo-oo!" Tucker is talking a blue streak right now. "Mom, we were best friends all summer! We slept in the barn together and grazed together and at night we walked up to the house together to visit you and David, and in the morning when you called us for breakfast, we galloped through the field right up to the barn and made you laugh every time! I thought she loved me, Mama. . . . I thought she loved me." I feel like my young friend is saying all this and more as he continues to wail.

"Mmmmm," he moans as he lowers his head, pressing into my thighs the way he loves to do.

I move the round, rubber currycomb in circles over Tucker's cheeks.

"Moo," he tells me.

"I know, Tucker. I didn't expect this either. I'm so sorry," I say.

"Mmmm," he continues.

"We'll figure something out," I promise, before heading to High Falls for Franklin's return trip.

～❯

Ninety minutes later, Franklin steps out of the trailer and walks into the spacious new pig pasture at Catskill Animal Sanctuary. With a Plan B available, we've decided to give the previously uneasy relationship between Franklin and his former roommate, Piggerty, another chance. Piggerty despises big pigs Jangles and Farfi, and is at best ignored by new arrivals Roscoe and Reggie. "She's been seriously depressed all summer," April explains. For my part, I hope Franklin's newfound confidence will serve him. If he stands up to Piggerty, there could be a happy truce. If not, we'll walk him across the road and into Plan B.

Franklin's head is low as he explores his old turf: he smells the pile of boulders, checks out the water trough, roots in the dirt. Piggerty is at the far end of the field; she has not seen him. Franklin crosses the creek, steps up the bank, and heads toward the beautiful new pig barn. I follow closely behind him.

"Look, Franklin, isn't this *gorgeous*?" I ask. Because it is. Franklin steps over the threshold and enters. For a moment, he simply sniffs the air, and then he looks at me, acknowledging the new space that smells of just-hewn wood and fresh timothy hay.

While Franklin explores his new digs, I step back out into the field and see Piggerty making her way toward the barn.

Allen is rounding the corner on the tractor, and as he pulls it over to park it under the willows, I call to him. "We might need you any minute now. Can you come in?"

Allen loves all the animals, like the rest of us, but he's mad about the pigs. He enters the field just as Franklin exits the barn; Piggerty is only thirty or so feet away. They have seen each other. There is no need for words here; Allen and I both know that this could go one of several ways, and we know what to do if the worst happens. (Animal introductions differ tremendously by species. In our experience, cow and sheep introductions are the easiest, followed by goats and horses. Rabbits, geese, ducks, and pigs are the most uncertain, complicated . . . even dangerous. One well-known sanctuary lost a pig during an unsuccessful introduction.)

Franklin and Piggerty are now snout to snout. Franklin talks. Piggerty is silent—not a good sign—but the playground bully seems happy to see her former mark. They dance around each other, then turn away. They circle together, Franklin on the inside, leaning into Piggerty's bulk. They walk side by side for twenty feet, each eyeing the other, and just as Allen and I say to each other, "I think it's going to work!" Piggerty pounces, digs into Franklin's flesh, and Franklin is running for his life.

Frantically Franklin darts behind the rock pile with Piggerty on his rear. Allen and I split up—I to cover Franklin, he to discourage Piggerty—but at the soonest possible instant I open the gate wide for Franklin's escape. For while I generally understand that most pigs need to be given room and time to work out their issues (we prefer they do this under supervision), Franklin is *always* the victim. Though at Summer Camp he learned to stand up forcefully to a horse and a cow, he's never stood up to a pig. With so many more living options for him than there were prior to his departure, I'm not prepared to risk his newly acquired confidence and security.

"Come on, pig," I say to my pal as I open the gate. "It's time for Plan B."

~~~

We introduce Franklin to the potbelly field directly across the road. Mama, Shy Girl, Muffin, Mabel, and Charlie live in this field with its spacious heated barn. With us for years, Ozzi is the newcomer in this group. Determined to use the entire farm as his buffet table, Ozzi was simply getting too fat; his free-range privileges were recently revoked. I sit and watch. Franklin is utterly uninterested in the potbellies; they are equally uninterested in him. The dominant Muffin makes no effort to confront Franklin, while Franklin simply settles in immediately, going right to work exploring his new digs. Fascinating—apparently, among the porcine set, size matters a great deal. Without exception, when we've introduced pigs to other pigs of similar size, the introductions have been epic: stressful, dramatic, tense. For Franklin, the little pigs may as well be pottery. He completely ignores them and goes about his business. They do the same.

These are the pigs with whom Franklin will sleep. The next piece of the plan involves Franklin sharing the huge adjoining field, where Hope has successfully integrated into a lovely herd of gentle horses. Tucker is now grazing on the edge of the herd; I suspect that within a day or two, he'll have worked his way in. But if the group is kind enough to accept Franklin, or at least to ignore him, then all three Summer Camp friends win: Hope with her newfound confidence can hold her own in the gentle group; Tucker, who's never shown much interest in other cows, will be right behind the barn, where it will be convenient to lavish him with the attention he craves; Franklin will also be nearby, with a huge field to explore, human friends with whom to interact, and maybe, just maybe, his Summer Camp bovine buddy to pal around with.

"Come on, Franklin," Abbie and I say in the morning as we open the gate wide for him to enter the horse pasture. Franklin loves to explore and wastes no time coming through the gate. The horses are clustered some two hundred feet away. Hope is in the middle; Noah and an old mare named Crystal are closest to us. Tucker is only thirty feet or so from the herd, resting under a small tree at the edge of the pond. I walk halfway toward the horses, fold my legs, and sit down to watch what unfolds; I'll stay until it's clear that I'm not needed.

The next half-hour passes uneventfully. Franklin does what pigs do: he snuffles and roots, walks a few feet, tears off a mouthful of grass, and snuffles some more. He barely lifts his head from the ground. The horses' heads lift, though, as one by one they spot him, the pink tank proceeding through the pasture, driven by his nose. But that's all they do: they see him, they look for a moment or two, then they put their heads down to graze. When Hope spots him, she looks for a moment longer than the others, but doesn't leave the herd to say hello to her summertime pal. Only Noah, currently the lone gelding in the pasture, looks for a good long while, then turns his head around to look at me.

"Is he supposed to be in here?" I think he's asking.

"Thank you, Noah!" I call to him, then move toward him to acknowledge his gesture with a good long head scratch. "Thanks, good boy. That's Franklin. He's a good, good pig. Thank you for telling me that he's in here. Thank you for sharing your pasture with him."

Other horses would terrorize pigs. But this is an uncommon group of generous spirits, and the introduction proceeds without incident.

Just a few yards from Tucker, Franklin either sees or smells his buddy, and in an instant is barreling toward him. "Mmmm-phh!" he calls as he approaches. "Hi Tucker!" he says.

Tucker licks him, remembering, as I, in heaven, remember the first time white cow muzzle met pink pig snout.

Plan B is not the perfect solution, but it's pretty damned close.

PART II

just
another day
at CAS

Nine
Happy Horses

My house is situated high on CAS property in the middle of a horse pasture. From my back deck I can see the four cow pastures, one sheep pasture, four of our eight horse pastures, the duck pond, our "special needs" area for blind duck, Sassafras, and his protector, Succotash, one of our rabbit houses, the turkey barn, the pig paddock, and three of our chicken houses. From my office window I can call out to our blind cow, Helen, and her devoted seeing-eye calf, Dozer, who, at 6:10 on a frigid April morning, are still snuggled in their barn.

Indeed, I'm a lucky woman.

The sky lightens slowly. Beyond the pond outside my back door, sixteen cows rest peacefully at the far side of their pasture. Only young Jesse stands near the pond. At three years old, he's still filled with wonder, and he's fascinated by the wild turkeys that strut and preen and peck the ground in front of him.

Athena, Fritz, and Abby stand at my deck, staring at the door. Six other horses—Mango, Mary Anne, Hazelnut, Callie, Eloise, Katydid—turn from the pond's edge to join their pals. Still in their fuzzy winter coats, the horses nonetheless look good, and I'm pleased. They've all gained at least two hundred pounds since their rescue from a Saratoga horse farm whose owner admittedly "just didn't want to feed them." I smile to myself, stunned that after just three days in this particular pasture the horses already have my number. They know that all they have to do is ask (and they're doing it beautifully by simply crowding the deck and *staring* at my door) and I'll emerge, treats in hand.

I take a five-pound bag of carrots from the fridge. "Good morning, girls and boy," I say. (Fritz was the lone boy from this particular animal rescue.) I sit on the deck and a pile of rescued horses surrounds me. I'm struck by their patience and politeness as they wait for me to open the bag and dole out the orange prizes. There's no jostling or competition—even Athena, the head honcho, allows others to cluster more closely than she. Beautiful Abby—pure white—nuzzles the top of my head as she waits. Abby is a wonderful success for Catskill Animal Sanctuary. Near death when she arrived, she could also barely walk: her hooves were a foot long and riddled with abscesses. It took over an hour for her to limp, one painful inch at a time, off the trailer when she arrived. But here she is just three months after her rescue, galloping from one end of her pasture to the next, no sign of pain.

The horses munch their treats as I tell them how fine they look. They're grateful to be here. While skeptics or cynics or obtuse scientists would say I'm being anthropomorphic, they'd be so dead wrong. Rescued animals show their gratitude in myriad ways obvious to anyone who knows and observes them.

I wait for them to saunter off. They'll head toward the barn, knowing that it's nearly time for breakfast. Sure enough, after a few minutes of pats and praises, Athena turns. She has heard April exiting the kitchen with the breakfast bowls. Eight other horses eagerly follow, and another day begins at Catskill Animal Sanctuary.

The Underfoot Family

In my lifetime, I've probably been in well over a hundred barns. My Dad's farm had five barns. I took riding lessons at many different farms throughout my childhood, and I often accompanied my Dad to various farms and racetracks. I've been inside the barns of many other rescue organizations, those of people wishing to adopt animals from us, and those from whom cruelty agents are seizing animals who are not being cared for. I've been inside plenty of barns that exuded misery and suffering, and thankfully, many that were filled with joy. But I've never been inside a barn like the main barn at Catskill Animal Sanctuary.

Take what's happening right now. I'm sitting in the middle of the barn aisle, trying to write about the Underfoot Family. Only I can't, or at least it's difficult, because Barbie the hen is insisting on getting in my lap, and her boyfriend Rambo, resident wise man, security guard, and the *real* boss of the joint, keeps pawing my thigh with his hoof because he wants a butt massage. Rambo is a sheep. If today is like most days, I'm sure he's already gotten, oh, at least a dozen butt rubs, but here I am, sitting in the aisle, another mark. (Obviously, I'm the boss only in theory, just as April and Karen are our theoretical farm managers and Abbie our theoretical animal care director. One doesn't need an organizational chart to understand the hierarchy here: it's plainly apparent who runs the show.)

I am being swarmed by animals. PeeWee the goat is nibbling my hair, and Hannah, Lumpy, and Aries, our other three free-range sheep, are lined up in front of me like three panhandlers. Charlie the pig walks past. He's on a food-finding mission, the mission shared by pigs worldwide.

Our newest free-ranger is a horse named Casey, the going-on-forty-years-old pinto gelding who arrived years ago after our concerned farrier told us of a horse living in a New Paltz junkyard. Casey, who survived the death of his best pal, Junior; Casey, who survived an abscess that swelled up like half a honeydew when he tore his cheek with too-long teeth; Casey, gentle giant, friend to all who call CAS home. Our low-lying pastures are the only ones suitable for Casey's old, arthritic joints, but in the spring, the mud is simply too hard on him.

So now, each morning, Casey's door in the middle of the barn is opened, and he heads out—sometimes going left, sometimes right, sometimes immediately to the hill in front of the barn to graze, other times to visit the horses in the large pasture behind the barn. Casey visits the cows; he naps under the willow tree next to a pig or two. After a particularly full day, he often settles in the middle of the barn aisle in the midst of the smaller free-range animals until his eyes grow heavy with sleep. Yesterday, he was relaxing in my yard, and if Murphy's old-dog ramp had been a bit wider, I've no doubt that he'd have walked right into my house. Casey delights in this life.

Why do we allow this? It's a model, after all, that doesn't exactly lend itself to efficiency. You try wheeling a cart filled with food dishes down the aisle and see how far you get, or try moving the tractor and manure spreader forward during stall cleaning when a sheep is standing in front of you and a chicken wanders between the front wheels and two pigs are arguing on the floor to your left. Before the driver can inch forward, she must wait for her assistant on the ground to shoo the gang out of the way and issue the "All clear!" signal.

Among the eight of us—April, Troy, Alex, Abbie, Karen, plus April's partner, Allen, and Alex's partner, Kathy, who fills in for Abbie on her days off—Troy is probably the best at setting limits. Alex loves to shout "Free the pigs!" as if a pig living in a huge heated barn with daytime access to a pasture that's even huger were somehow imprisoned. If Allen had his way, every animal on the farm would be free to wander when and where he chose, and at least once a week, Abbie has added another being to our growing roster of free-range animals. Truth be told, even Troy, Mr. Efficiency, revels in the freedom we give the animals who need it. Every afternoon, Troy hops on our bike in search of the missing Wilbur. He's a pink potbelly pig, and of all the roamers, he ventures farthest from the barn. One never knows where he'll be at feeding time. Sometimes it takes Troy two minutes to find the little rascal, sometimes ten, but eventually what we see is always the same: a little pink pig *sprinting* back to the barn on four-inch legs, followed closely by Troy, pedaling furiously as the pair charges through the horse pasture, reminiscent of the Wicked Witch of the West in that memorable scene with Toto.

Its inconvenience to humans notwithstanding, I've always insisted on this model and seem to draw people who revel in it.

The belief that every living thing deserves happiness, that all hearts deserve to sing, informs all that we do at Catskill Animal Sanctuary. For as long as he or she is with us, we do what we can so that every single animal in our charge can thrive. And for a host of reasons, the quality of care we provide to most—large spacious pastures, generous barns and shelters, a superb diet, friends of one's species, and plenty of attention from humans—isn't enough for all. Blind animals, old animals with a variety of old-age afflictions (most typically arthritis), timid animals who need more frequent attention, animals who are being picked on by their herd or flock because they're old or small or timid or all three—these are the ones who join the motley crew known affectionately as the Underfoot Family. If there remain in any of us vestiges of ignorance or species-specific prejudice (for example, sheep are shy, turkeys are stupid), these animals are in our faces each day to challenge those prejudices. And regarding anthropomorphism, if you buy that concept (which I do not), if you believe that this entire book is anthropomorphic, I challenge you to come to Catskill Animal Sanctuary, to sit quietly for half an hour among various members of the Underfoot Family, and to tell me at the end that we're attributing "human" emotions to animals.

Right now, for instance, it's extremely difficult to type. Why? As I sit in the middle of the barn aisle, laptop on my lap, our free-range turkey, Ethel, is lying next to me, her feathery body pressed into my side. As I attempt to type, so does she. She's enthusiastically pecking the keyboard.

"Oh, I see you've got a few assistants!" Karen offers as she walks out from the kitchen, her arms stacked with feed dishes for some of the smaller animals.

"I don't need your help, Ethel," I say to her; evidently her hearing is selective, because she continues typing. Across the aisle from us, Zoey and Charlie, two potbellied pigs, are arguing fero-ciously over who's going to get the favored spot in their enormous bed of hay: both always want to be closest to the wall. Meanwhile, Hannah of the Three Wise Sheep moves forward until she is two inches from a turkey named Agent Forty-Four, at which point Aries and Lumpy turn and head outside to discover what awaits them.

I was going to make a point about the closeness of this family, about the seamlessness of their days, about what a privilege it is to observe their adventures, their kindnesses, their unequivocal accep-tance of each other, their schemes (all of which have to do with breaking into the kitchen). I wanted to discuss how much love they give to us, and to each other, many times a day. Only, again, the Underfoot Family is making this extremely difficult, as Casey the horse has just wandered in from his visit with the cows and has joined Hannah, Ethel, and Petunia the pig, who is pressing into my knee, then my thigh, with her snout. So here I am, surrounded by animal friends who could be anywhere they choose but have all chosen to join me in the middle of the barn and assist in the writing of this chapter. And though I do apologize for needing to end it so abruptly, I trust that you get the picture.

A Girl Named Norman

L ike millions of turkeys every year, Norman was destined for the Thanksgiving table.

But thanks to an interesting twist of fate, some well-timed phone calls, and a few soft hearts, Norman celebrated Thanksgiving with the rest of the crew at Catskill Animal Sanctuary.

When WSPK, a Beacon-based radio station, advertised "turkey bowling" in its parking lot, the calls and e-mails, all of the "You've *got* to save the turkeys!" variety, poured into our office.

Surely the event was a prank. Curious and concerned, though, my assistant Julie and I drove down to Beacon, video camera tucked beneath Julie's arm. (Many, many thanks to my "Anything for the animals!" Julie, who woke up way, way too early for her liking.) There, though the on-air DJ bragged about "the crowd," a mere seven spectators stood in the cold, waiting to bowl *frozen* turkeys at ten pins borrowed from a nearby bowling alley.

A blue SUV pulled up with "Norman," the frightened turkey borrowed from a nearby turkey grower, in the back. Evidently, as if this entire event weren't nonsense enough, the station also needed a live prop.

"Who's gonna get him out?" a heavy woman with heavier makeup asked.

"It's a *she*, not a *he*, and you need to be careful. She's already terrified," I said. Julie stood just behind us, recording the spectacle.

A guy reached in, grabbing Norman by the wing.

"That is *not* how to hold a bird," I stated flatly. "Wrap your arms around her so you can pin her wings and support her weight. Otherwise, you'll both get hurt."

Norman and her cage were set up between two loudspeakers. The DJ continued to spin the story, describing how the turkey was having fun, the crowd was having fun . . . golly gee weren't we all having *fun?* Meanwhile, all Julie and I saw were a terrified bird, seven cold people, and three Butterball carcasses waiting to be slid down a "bowling alley" composed of plastic garbage bags.

⚓

So there we stood in the parking lot—twelve people if you included Julie, me, and the radio staff—and a frightened, hyperventilating turkey locked in a crate.

"I'd like to take this bird to Catskill Animal Sanctuary," I said to Jason, the station manager, who had apparently been called outside because two strange women were much more interested in the *live* turkey than they were in the "turkey bowl" competition.

"He's not ours," Jason explained. "He's the property of Quattro's Poultry Farm. And what's Catskill Animal Sanctuary?"

I explained that Catskill Animal Sanctuary was a haven for abused farm animals, and that this was clearly an abusive situation. "And the turkey is a *she*, by the way."

"Look," he said, his eyes dropping. "I just wanted people to have fun. It's a holiday. It's supposed to be festive."

I softened a little. "Does it look like they're having fun? You've got only seven people here, and three of them look like unless they *win* the competition, they won't be having Thanksgiving dinner." I motioned to a chain-smoking mother and her two gaunt young daughters, all noticeably underdressed on this frigid day.

Jason hesitated but a second before giving me Quattro's phone number. "What they want to do with the turkey is their business."

"Thanks," I smiled. "You know, you might rethink this event for next year. You're welcome to come celebrate at Catskill Animal Sanctuary—people *will* have fun . . . and so will the animals."

"Sounds good," he said.

∿

"Sure, you can have the bird," said Carmen, evidently the owner of Quattro's, over the phone. "You'll have to buy it."

For many reasons, Catskill Animal Sanctuary does not advocate purchasing animals in order to retrieve them from desperate situations. While we're contacted routinely to save animals on their way to auction, for instance, we generally decline to intervene. Many sanctuaries draw an even harder line than we do—absolutely no purchasing of animals, ever. Why put money into the hands of someone who will simply purchase more animals to abuse? That's how the thinking goes: through one's purchase, one continues the cycle of mistreatment.

But this was an exceptional situation. "Norman" had some degree of notoriety, as the radio station had been hyping its "turkey bowl" for weeks. If she could bring guests to Catskill Animal Sanctuary to discover that turkeys, cows, pigs, chickens, and other animals that most humans eat are remarkable in their own right, then we needed to find a loophole in our "no purchase" policy.

Julie and I pointed the car in the direction of Pleasant Valley.

FRESH KILLED CHICKENS read a huge sign on the porch of Quattro's old clapboard general store. I stepped inside. A line of people waited at the single cash register. Each person held a newly slaughtered turkey. Some had geese, ducks, and pheasants as well. At the back of the store, guns, ammunition, and camouflage gear lined the shelves.

"Hi," I said to the cashier. "I'm looking for Carmen."

"She's at the counter," she said, pointing behind her.

Another long line. It was, after all, the day before Thanksgiving, and this was *the* place, apparently, if you wanted "fresh killed birds."

A man weighing easily five hundred pounds hoisted each package to its eager recipient, who then proceeded to the cash register.

I approached him. "Is Carmen here?"

An elderly woman walked toward me. "Kathy?"

"Yes. Hi, Carmen."

She was a small, bent woman, around eighty years old. Though her hands were gnarled with arthritis, they were strong hands. Carmen was a worker.

She came toward me and took my hand, pulling me to a screen door. We walked into a pantry, away from the eyes and ears of her employees. She looked up at me. "I love animals," she whispered. "I love all animals. I love these birds. I wouldn't do this if I didn't have to."

I could have said so much in that moment, but instead said only, "Why don't you come visit Catskill Animal Sanctuary?"

"Yes. I'd like to do that."

I went to the car to retrieve a brochure, and on it wrote my name and phone number.

Carmen returned to her place behind the counter. I walked out, hurting not just for the millions of birds senselessly slaughtered for this one holiday, but also, somehow, for the person responsible for many of those deaths.

~~~

At the bottom of the drive, a stressed-out Norman paced in the same cage that had held her between the radio station's speakers.

"Hi. We're here to pick her up," I explained to a toothless gentleman who approached our car.

"I'll get her for ya," he offered, and before I could span the few steps between the car and the turkey, he reached in to drag her out by the feet.

"Please, let me do it," I insisted as I pushed myself between him and Norman, her terror rising again. I squeezed my upper body into the crate, hovered over her, pressing my chest into her back and wrapping my arms around her sides.

"It's okay, girl," I whispered to her as I cupped her little head to get her safely from the crate. "You're safe now. You're going to a wonderful place."

Julie and I settled Norman into the rear of my Subaru wagon in a large crate thick with straw, then began the slow drive back to her new home at Catskill Animal Sanctuary. There, on the day for which millions are slaughtered while those responsible give thanks, one single turkey will discover what it's like to have room to explore, a spacious barn in which to settle down each night, and the freedom to determine how she spends her days. She will be the one to decide how much interaction she wants with other animals—turkeys, pigs, sheep—the various members of our free-range family, while one after the other, her friends face the death squad.

～⌒

One of the most striking and delightful discoveries in the seven years of doing this work is how much individuality there is within each species. I'm sure I knew this intuitively, but to experience it is something altogether different. It's yet another way that animals are similar to, not different from, their exploiters. At Catskill Animal Sanctuary, this issue matters to us. Inasmuch as we can encourage each animal to become "who she is," it maximizes the animal's happiness and gives us more ammunition. The more our animals are allowed to evolve, the more effective ambassadors they can be for all members of their species. The common emphasis on the suffering of our food animals doesn't reach the person who doesn't care about cows or pigs, turkeys or chickens.

The prevailing view, of course, holds that our species is entitled to do what it pleases *to or with all other species*: might makes right. In the case of food animals, we're *entitled* to grow them for food—and to torture them throughout that process in the interest of profitability. We're entitled to wear their skin. We're entitled to profit from all their inedible or unwearable parts by grinding them up, boiling them, adding chemical and color and scent and more and turning them into food for other animals, personal care products for humans, and more. Many, most notably Dr. Will Tuttle in his important work *The World Peace Diet,* argue that this unthinking, wholesale, systemic violence is the root of all violence in the world, and that until we as a species recognize that fact, the world will never know peace.

I'm not thinking about this now, however. Today, we've invited Norma Jean (as Norman has been renamed) to join the Underfoot Family, as she is, for the moment, our lone turkey, and I'm having way too much fun watching her discover this big new world. Indeed, it is always a privilege to watch each new animal make the choices that so many others have before her: how far to venture out from the safety and certainty of the barn, whether to seek out human friendship or keep her distance, which animals to befriend. Will she become a lap turkey, or will she keep her distance?

One day it becomes clear that a new goat will never be accepted by the herd. A few weeks later, a pig arrives who's spent most of his recent life in a crate. Then a young calf, slaughter-bound but for good fortune, who's way too young to be introduced to either of our cow herds. The reason varies, the goal stays the same: we want every animal to thrive. Those who need the company of others of their species join their respective flocks or herds in their respective roomy pastures. But there are dozens who fall into the "special needs" category and they are assured a spot in the main barn.

"Come on, girl," I say quietly as I open her stall door wide. "Come meet all these new friends!"

Norma Jean's only world to date has been filled with birds like her. Turkeys and indifferent (or worse) humans, more than likely, are all she has known. No surprise then that it takes her over ten minutes to gather her courage, but finally, here she is, standing in the threshold. Her round eyes blink. She is motionless except for her head, which moves in every direction to take in the big world before her. Charlie the pig grumps per usual about the lousy state of affairs as he walks past, uninterested. Our vice president and long-term volunteer Chris Seeholzer moves from stall to stall, filling water buckets, whispering sweetly to each animal as she does so. April enters the far end of the barn with Big Ted after the old draft horse has spent the morning outside; she speaks gently to him as they walk. Just feet from Norma Jean, Jack, an ancient, blind sheep, is playing with a broom hanging on the wall. I imagine our feathered friend is thinking something along the lines of, "Toto, we're not in Kansas anymore."

~~~

Just three days after Norma Jean joined the Underfoot Family, I walked into the barn to discover her lying right next to Rambo the sheep, her feathered body pressed into his wooly one. Rambo, once the most violent and dangerous animal I've ever known, allowed this—even allowed Norma Jean to peck at his wool, pulling at whatever bits of hay she found interesting. How quickly he became her source of comfort and security, as he has been to so many others who somehow sense his energy, his strength, his empathy. We call Rambo our guard sheep, but he's so much more than that. For us humans, he's the consummate teacher whose many acts of bravery, wisdom, and empathy demand that we throw out our assumptions about the supposed differences between humans and animals. The animals look to him as leader.

Rambo has never sought out friendship. Rather, over the years, one animal after another—many sheep, a goat, a duck named Peepers, a potbellied pig, several chickens, and now a turkey—approached him. They need him; there's simply no other way to explain what we witness, for all Rambo does is lay on his deep bed of shavings and hay against one of the barn walls, and the animals come to him. Like so many others, Norma Jean's favorite spot in the barn is wherever Rambo is. In the morning, she scrunches up right next to him, pressing her bulk into his body. When Rambo gets up to stretch his legs or take a stroll, Normal Jean is never far behind. A few years ago, I'd have thought, "What an unusual friendship!" Not anymore. Rambo is always the go-to guy.

"My tom turkey needs a wife," Patty Reller said when she called us about another adoption. Patty adopted ducks from us several years ago; her small "Hollyhock Farm" should be renamed "Hollyhock Heaven." Patty completed another adoption application, and a few weeks after she joined us, Norma Jean left the Underfoot Family at CAS to join the underfoot family at Patty's place.

In addition to dogs Snorty, Bean, and Hazel, Patty shares her home with four cats, several parrots, an impressive variety of chickens, Flubby the horse, ducks Charlotte and Sir Francis Drake (Charlotte was named after Charlotte Mollo, who serves with her husband Walter McGrath as our dedicated adoption screener), three donkeys, and a delightful goat named Captain Hook.

I enter Hollyhock for a routine check-up after a long drive up a beautiful country road and am greeted by curious animals delighted that someone new is paying them a visit. But trying to take notes as one visits is like taking notes in our barn: Daisy the donkey alternately grabs my pen, my notebook, my sleeve, my hair, and doesn't let go. When I back up to put some distance between us, she follows me. Naturally, I could get out of the field and simply observe from the driveway, but then both Daisy and I would miss out on the fun.

"Norma Jean is a widow," Patty says to me with a straight face, describing how she had lost "both her husbands." But she goes on to speculate that the gentle, unassuming bird is probably happier with Mabel, her female friend, since both her "husbands" were a little aggressive.

It is a cold day when I visit Patty, Norma Jean, and the menagerie who call Hollyhock home. Norma is in her barn under a heat lamp, no more than a foot from her pal Mabel. Like the rest of the birds, cats, and dogs, she's free to roam wherever she pleases on this little piece of heaven. She didn't remain much of a lap turkey, as Patty describes her as "a little reserved." Still, this one turkey really is living happily ever after. How I wish their lives mattered to the people three deep at Quattro's, buying "fresh killed birds" the day before Thanksgiving.

The Audacity of Love

That Hannah the sheep is in love with Rambo the sheep is no secret. Indeed, it's obvious even to first-time volunteers as Hannah bolts from her stall each morning in search of her Romeo. If she finds him immediately, all is well. But if Rambo is out of sight—either intentionally hiding or simply munching hay in a newly-vacated stall—she is initially disturbed, then worried, and finally panic-stricken, uttering a heart-wrenching, baleful *baa-aah* as the time it takes to find her soul mate increases. Once she locates him, all is again right in her world. She settles into her sheepness, content to roam the barnyard, grazing, stealing alfalfa from the hay room, and plotting kitchen break-ins . . . that is, as long as Rambo is no more than a foot or two from her. It is a relationship that she needs desperately, and one that Rambo sometimes seems to appreciate, other times only tolerate.

Enter the other woman.

Barbie is a "broiler," the term used by the poultry industry to describe chickens intended for meat. Broiler: they exist for the sole purpose of being broiled. Or baked. Or barbequed. She's one of hundreds who've arrived at CAS over the years from one of New York's five boroughs, lucky escapees from live poultry markets, slaughterhouses, transport trucks, and the ritual sacrifices of Santeria. We've taken chickens from dumpsters, chickens tied to trees in Central Park, chickens stuffed in mailboxes, and chickens who were drowning in crates left in flooding streets. Our latest, Barbie, was found in Brooklyn, hiding under a blue Honda.

Like Hannah, Rambo, and many more of our smaller animals, Barbie free ranges during the day. While she is young, the exercise is good for a body that will quickly grow morbidly obese. There's also no outdoor home for Barbie, as our ratio of roosters to hens is about 300 quidzillion to one. (Few people, unfortunately, want a pet rooster.) So Barbie snuggles into her home in the main barn each night, then each morning is lifted out to explore the barnyard and cozy up to whomever she chooses.

Unfortunately, just like the rest of the girls who've come before her, Barbie has chosen Rambo.

For several weeks, Barbie has been napping right next to Rambo, sometimes so close to him that surely even through his wool Rambo feels the heat emanating from her big bird body. Sometimes she climbs on top of his back, the patient Rambo motionless, and falls sound asleep. Rambo takes her overtures in good stride.

For a while, Hannah tolerated the new friendship. After all, Barbie was merely a hen; Hannah could still rest side by side with her love, or stalk him relentlessly as he traveled the barnyard ensuring all was in order.

But Rambo, the most exceptional animal I've ever known, had other things in mind.

A couple weeks ago, I stood, incredulous, as Rambo walked up to Barbie and pawed the ground. Pawing is Rambo's signal to humans that he wants a massage—something he receives whenever he asks for it, which is generally, oh, forty or fifty times a day. Clearly he thought that if human beings could discern his wishes, then a chicken could, too. We stood there, my hand on April's arm, both of us gaping, as our extraordinary friend tried to teach his bird pal to do his bidding. When it didn't work, Rambo finally took the tip of his horn and very gently massaged the little bird.

A few days later, Rambo was lying in a pile of hay. Next to him was Barbie, pulling bits of hay from his wooly coat.

The deepening of this relationship was too much for Hannah. One recent afternoon, she was nowhere to be found as I entered the barn to set up feed.

"Where's Hannah?" I asked Abbie.

"She's in time out."

"What happened?" I asked, imagining her response.

"She head-butted Barbie halfway across the aisle."

Tension mounted when Barbie began using Rambo as a sofa. At seventeen pounds, she is no longer fully mobile, yet she can still manage to climb atop the resting sheep, take a moment to decide whether she wants to face his horns or his woolly rear, and ease down into fluffy bliss.

For weeks I'd heard about (but not witnessed) this new development in the relationship between the great sheep and the presumptuous chicken. And then one crisp December morning,

I exited the feed room, and there, in the middle of the aisle, were Rambo and Barbie. Barbie was one happy hen plopped dead center onto Rambo's back; Rambo was completely unfazed. I dashed back inside for a camera—people *had* to see this—and then moved slowly toward them.

"Rambo, you are a prince," I praised him. "What a kind man you are," I said as I sat down just feet from them to snap the best shot.

And then I heard it: the rapid click-click-click of sheep hooves moving toward us. It was Hannah. She had spotted them.

The ball of brown wool pushed past me as if I weren't there and strode within six inches of the offending pair, neither of whom budged. She glared at them; she looked at me. She looked back at them; she looked at me. There was no need for words here, as *Are you going to help me here, or what?* or really, most precisely, *What the hell is this?* were etched into every gesture.

"I'm sorry, Hannah," I whispered, approaching her with consoling words. But Hannah pooped and marched outside, wanting nothing more of Rambo, the interloper, or me.

⁓

I never imagined I'd work at a place where a sheep and a hen would vie for a second sheep's affection. But then again, I never imagined that a dying cow would lick my face over and over and over again until he took his final breath, or that a former cock-fighting rooster would evolve into a being who begged us to share our lunches, took car rides with me, and happily climbed onto my dog Murphy's bed to share a nap.

These are the things that love allows. Animals are far more like us than I'd have ever imagined had I not had the good fortune to be with them every single day of this most wonderful life.

If We Can Get Them Out

In the midst of this country's economic downturn, it has been our pleasure and our privilege to receive animals from caring people who reach out for help because they can no longer afford to care for their beloved family members. Farms are in foreclosure; people are losing their rural rental property . . . or simply can't afford the price of hay. They call, generally in tears, when they have run out of options. We help as many as we can.

This situation was different.

I was called by another rescue about a woman who "needed help." She had three animals—a mare and two older stallions, all Thoroughbreds—and could not afford them. "They might be in bad shape," the rescuer said. State police were weighing whether to arrest the woman or simply to force her to surrender the animals in order to avoid prosecution.

I called the woman. She talked about how she'd lost her job, how she could barely afford to put gas in her car, how the horses were forty minutes from her home, how she had to haul hay and water in a broken down vehicle that might just die on any of these trips, and how no one wanted the animals—especially not two older stallions.

"They're Thoroughbred stallions," she said. "Who in their right mind would take them?" I refrained from asking why she didn't neuter them many years ago. (In conversations with people who aren't taking care of their animals and who *seem* willing to surrender them, it's essential to remain unruffled. A comment made in anger or frustration can easily derail the process.)

We would take the stallions. Despite all the challenges— needing to house them far away from any mares, not being able to turn them out with any other horses, the expense of gelding, the risks associated with gelding older horses (one is fourteen, the other twenty-one), and the likelihood that they will be extremely hard to handle—we would take them.

I told the owner that we needed to see the boys before they came to Catskill Animal Sanctuary, and that we would find a good home for the mare. Questions loomed. Were they strong enough to make the journey to their new home? A psychologically damaged stallion could be extremely dangerous; what had their imprisonment done to them? What was their physical condition? If they were exceptionally thin and debilitated, we'd first need to put weight on them before they'd be able to withstand being gelded, and that would mean more time juggling the management challenges inherent with stallions.

April's partner, Allen Landes, who works full-time as a hospital biologist at Albany Medical Center but is nonetheless a godsend at CAS every Thursday, lives just three miles from the horses' home.

He agreed to take a look at them. I relayed directions from the owner; Allen went right after work.

He used his cell to call from what he described as a long-abandoned, derelict camp of some sort.

"Kathy, this can't be the place," Allen said. "There's no sign of life here. Are you sure you understood the directions?"

"It's the place," I assured him, having gotten a clear description from the horse owner.

Allen walked into a place that looked like it had not seen life in decades. "No horses in here," he said as he traipsed through one ramshackle shack after another. "Nobody here," he said as he moved through the derelict remains of another. Twice more he asked whether the directions were accurate.

Allen was about to give up when he spotted a forlorn-looking barn at the base of a hill at the back of the property. "Ugh," he uttered. "I hope they're not in here."

They *were* "in here."

For many years, Allen served as a board member for an Albany-based rescue and assisted with seizures when owners were arrested for cruelty. The first words out of his mouth were, "God . . . this is the worst situation I've ever seen."

Three horses. All bone thin. Living in darkness in tiny, rat-infested stalls that had never been cleaned. Blanketed, but with blankets that were stuck to their manure-encrusted coats and would have to be cut away from the animals' bodies. Water buckets bone dry. "There's no sign that there's ever been water here," Allen said. One stallion was shivering violently and spent most of Allen's half-hour or so there lying down, rats crawling around him. His breath was rapid and shallow.

It turns out that the woman had been arrested for animal cruelty eight years ago, when one of her horses, a Thoroughbred stallion, had to be euthanized—he was simply too far gone. We realized that getting those horses out would be the most humane option.

~

On moving day, Allen and his friend Bob met Corrine the hauler, Sue McDonough—a thirty-year veteran of the state police force and a crackerjack cruelty investigator—and Tina Murray of Harmony Hill Rescue. The horses' owner met them at the property; Sue would determine later whether to file formal charges. For now, the group's only goal was the safe removal of the animals.

The ramshackle, windowless building that had been their prison for so many years was at the bottom of a steep slope. Corinne had to park some hundred feet from it. The group would walk the horses out to assess their condition, and, they hoped, successfully load them.

"The mare came out first," Allen told us later. And that was when the owner made the first comment that nearly sent him over the edge. Her blanket was not only filthy and encrusted, it was so torn and tangled that traveling in it would have been dangerous. As slowly and carefully as they removed it, however, they couldn't avoid taking much of the mare's skin with it, and when it finally came off, the woman eyed the bald and scabby body, some two hundred pounds underweight, and said, "Oh, I thought she was going to be skinny—she looks pretty good!"

Cas, the bright red stallion, was rail thin and shivering violently. "He was a wreck," Allen said. "It took two doses of tranquilizer before we could get him on the trailer." Eventually, under the hazy fog of Acepromazine, they did.

It was Noah, the final horse, who struggled most. The one that the owner referred to as her "boy," Noah left the barn taking baby steps; it was all he could manage on malformed, painful hooves. Recalling Noah's effort to leave his dark prison, Allen remembers two things: first, how Noah, accustomed to living in total darkness, kept blinking as his eyes adjusted to a sunny day; second, how Noah kept falling. "He was in so much pain," Allen said, "and he was just so weak."

Corinne's trailer that day was a "step up" with no ramp for the animals. The trailer floor was about a foot above the ground. The first two horses had managed to step up; Noah simply lacked the physical strength.

"He tried so hard," Allen said. He'd step onto the trailer with his right leg, then fall as he tried to bring his left leg up. Sometimes he'd crumple to the ground; sometimes he'd fall hard. At one point, the owner said, "Maybe we should leave him here." Allen later told us, "That was when I saw the steam coming out of Tina's ears."

But the old stallion did not give up, and neither did the intrepid team of caretakers. Eventually, they figured out how to sort of slide him onto the trailer with Tina pulling from the front, the others shoving from the rear, and Noah doing his damndest to help.

They had done it, and were on their way home.

❧

"Where are you?" I asked Corrine. The mare had been delivered to Harmony Hill; the entire CAS crew had been anxiously awaiting the arrival of the two stallions, and it was now nearly 7 PM.

"I'm on Old Stage," she said. "It's been a hell of a day."

I hung up and shouted into the kitchen, "They'll be here in two minutes, guys!" Six bodies bundled against the cold appeared.

"What kind of shape are they in?" April asked.

"Corrine says 'awful.'"

A tall, narrow chestnut was the first off the trailer. He hobbled one painful, unsteady step at a time as a chorus of voices welcomed him, praising every step. A torn and tangled horse blanket hung as stiff as cardboard on his narrow body. He was shaking violently—the most violent shiver I've ever seen in an animal. Whether the shiver was from cold or pain or both, we couldn't yet be sure. His eyes were uncertain.

"This is Cas," Allen told us. Evidently it was short for Casanova. Interesting, I noted to myself . . . his name was the same as ours.

Cas was a true chestnut—a bright rusty color—and as narrow as a reed. He was all bone, no flesh. He rocked from side to side as we cut the fetid blanket from his body, tightening and watching us warily.

"Can you get him something warm?" I asked April.

"What size?" she asked.

"A seventy-four would work fine," I responded, and April returned with a blanket that looked like it could melt an igloo.

She held Cas while I slipped the blanket over the pockmarked body. The hair on a quarter of his body was missing. When I rubbed my hand over what remained, it came out in hunks. Cas flinched; I stopped. "Okay, good boy," I praised him. "It's okay. You just eat and get warm."

A darker horse—a rare deep rusty-brown known as liver chestnut—was next. This was Noah. Noah had turned himself sideways during the trip, and as Walt attempted to turn him around, he collapsed once again, crashing hard to his knees. April, Allen, Alex, Kathy, and I crowded around the outside of the trailer, encouraging him.

"You can do it, Noah," I offered.

"It's okay, boy, you're okay," Allen said, and then to us added, "He's worked really hard to get here."

Walt steadied him and he stood. We watched as, inch by inch, he painfully made his way toward smiling faces. The mere act of walking was an act of bravery. I looked down and understood why. Noah's hooves were grotesquely malformed. His left front foot, in particular, was violently twisted, resembling a wrung-out dishtowel. Noah had fallen again during the trip. "He fell hard," Allen said. He was utterly exhausted. Yet still he moved toward us, one tiny, excruciating step at a time, until a few moments later, Noah was in what I hoped to hell felt like heaven: a deeply bedded stall with two buckets of fresh water and a hay rack filled to overflowing with the stuff of life.

When Cas and Noah arrived tonight, they reminded me: *this is why you do this work.* Yes, absolutely. Rescues like this are precious moments at Catskill Animal Sanctuary. They are, indeed, our moments of glory.

Over the next few weeks, visitors and volunteers meet these two boys. Many cry; many ask, "Aren't you angry?" In fact, judging by how often I'm asked the question, many people have a hard time grasping why I don't walk around in a constant state of rage. But I don't. I walk around with a heart so filled with joy that I feel it could burst at any moment. Anger serves neither me nor the animals we are here to assist. But joy does. Joy works.

～

Noah and Cas have only been with us for two weeks, and a woman I'll call Marsha is hounding me to euthanize Noah. Her phone calls are inappropriate and offensive. Her suggestion is premature at best; groundless, or something far more cynical, at worst. She saw Noah the day after his arrival at Catskill Animal Sanctuary, the day after he tried far harder than a lesser spirit would have to wrest himself from misery. She heard the stories of how he fell, time after time, of how he wouldn't give up. She heard about when he arrived at our barn and fell again, about how he took fifty or more miniscule steps between the trailer and his stall just twelve feet from the door. A horse who walked normally would have covered the distance in five or six steps.

"What are you doing about Noah?" she asks in her third phone call in two weeks.

"We didn't save his life in order to end it," I said to her. "We've got to see how he recovers."

I understand her perspective: she believes Noah is too far gone to have any quality of life, and that putting him down might be the kindest thing to do for him. It would also open a space for another needy animal. She doesn't say this; she knows she doesn't need to, but it's the real reason she's almost *lobbying* for death by lethal injection. She knows I know this. Marsha has been in animal rescue for a very long time; where we disagree is always about euthanasia. Her "just put him down" statements leave me feeling like she truly doesn't understand our mission. "Emergency rescue" does not mean "remove animals from wretched conditions, then after that if you need to euthanize them it's no big deal." To Catskill Animal Sanctuary, the definition of emergency rescue is something like this:

Always take the neediest first—the ones who have truly run out of options. If there is hope, no matter their physical or mental condition, *try mightily* to save them. If they need weight, give them the optimum weight-gain program for their species, age, and condition. If they need love, let them guide you in how you offer it: many aren't ready right away for a full-blown love fest. If they need encouragement, offer it in abundance. Only when there is no hope, but instead only suffering, make the choice that they would ask you to make if they could speak. Make it quickly and with confidence—prolonged suffering is not an option—and send them off with all the love you can muster.

It's true that when we look at Noah's gnarled hooves, it's hard to imagine that the coffin bone, the large bone inside the hoof, hasn't rotated. If it *has* rotated, then Noah is permanently crippled, and we'll have to wrestle with the agonizing choice of whether to have him live his entire life in a deeply bedded stall or whether instead to put him down. Most horses would become stir-crazy and depressed living in a small box; the decision would be unambiguous. I'm not so sure about Noah. He's a mellow fellow.

On the other hand, if by some miracle there is no rotation, will our farrier Corey Hedderman be able to reshape his hooves over time to prevent the constant pain that plagues him? Will Noah ever be able to walk normally? If not normally, then at least without pain? These are the questions being raised while we're in the "wait and see" stage of his recovery.

Yes, it may be that we ultimately do have to euthanize him, but while we're waiting to look at hoof and lower-leg X-rays, we are certainly not thinking about euthanasia. He's in pain, but not in agony, and he's far from giving up. Give me a break! Noah's attitude and appetite are superb; he's gained sixty pounds; the hard clumps of crusted manure have been removed from his coat as volunteer after volunteer asks to groom him. For now, at least, he appears perfectly happy to remain in his stall, his head always buried in his hay, while his strength returns.

No. We will not euthanize him.

"This is inappropriate, Marsha," I say before hanging up. "Please don't call me again."

❧

Once again, Noah's head was where it always is: buried in his hay. Even after a night of eating with gusto, I knew he'd be ready for more, so I entered his stall carrying two big flakes of timothy-alfalfa mix.

"Good morning, lovely boy," I said to him as I placed one flake, then two, in the rack on the wall.

Noah didn't want his hay in the rack. The old horse wanted it on the floor, which he indicated by lowering his head and looking directly at the ground. So I placed it there, in the corner. Noah tore out a huge mouthful, then lifted his head, looked right into my eyes, and blinked slowly.

Aaah, the blink that says so much, all of it good. You've seen it in your cats. All animals (at least the species I've been around) use it: slow, purposeful, direct. I believe it means everything from "I love you" to "Thanks for the food" to "I'm happy to be here."

I blinked back and said, "You're welcome, big boy. We're so happy you're here." In fact, I was instantly bliss on two legs, for that simple, slow blink told me a great deal. In essence, it was a thank you: an acknowledgement of what we're doing for him. Few rescued animals are willing to communicate so openly and so quickly. In fact for many, any eye contact whatsoever is far too threatening. But not for Noah. Despite tremendous and prolonged suffering, Noah has a big, open, grateful heart.

And like Rambo and Murphy, he's a communicator of the first order. "My foot hurts," he says, as he lifts his hoof when he knows we're looking, then often turns his head to ensure we see what he's doing. "Can you help me?" Noah speaks with his eyes; he speaks with slow turns of his head to look all the way behind his body. Like the great sheep Rambo, whose story is told in my first book, and the great dog Murphy, he's wise enough to know that we're paying attention, wiser still to find ways to tell us what's on his mind

~

"Today's a big day, beautiful," I said to him. "We're going for a walk."

The momentous decision to take Noah for short daily walks came the day I entered the barn and saw our farrier Corey standing in the aisle with a huge grin on his face. *Could it be?* I wondered.

"There's no rotation," he said. He'd just examined the X-rays with our vet Heather O'Leary.

"Get *out!*" I shrieked. Rambo startled; Barbie the hen ran for her life. We were all incredulous. As mangled and deformed as his left front hoof was, we were all but certain that the bone inside that hoof had rotated. But it hadn't. We had our shot. If Corey could work yet another miracle, Noah had a chance at a relatively normal life.

⁓

"Come on, big boy," I say as I slide green halter over red head. "We're going for a walk! Can you believe it?"

I take the end of the lead rope and move backward toward the door so that the rope's six-foot length extends between us. I figure it's better for Noah to navigate his own turn. He, not I, knows which joints, which hooves hurt, and how best to place them to minimize the pain.

At the open stall door, he stretches his head out, turning left and looking down the long aisle; turning right as Millie the potbelly pig trots past in her relentless search for food. I do not rush this process. Let each animal heal in his own way, on his own terms. That's the Catskill Animal Sanctuary way. While Noah's reluctance to leave could be *fear-based*—his entire world prior to coming here was a dark, windowless, ramshackle barn—I strongly suspect this behavior is *all* about a reluctance to leave his hay. Like many chronically starved animals, Noah is obsessed with food. Perhaps he needs a few moments to get comfortable with the idea of leaving it.

It's been three or four minutes of standing, looking, but here he comes toward me now. I want him to see the encouragement in my face, and I want to be able to watch how he's moving. So I walk backward, facing him. Noah moves cautiously, and in slow motion. He steps tentatively with his front feet; I notice his back legs drop dramatically at the fetlock joint—he has the equine equivalent of extremely weak ankles. Noah walks about thirty feet. Then he stops, looks at me, turns back, and looks at his stall.

"Okay, babe, enough for today? Thanks for telling me," I praise my friend. We circle as widely as we can, and take more small, cautious steps back to the comfort of his stall. Within seconds, Noah's head is buried deeply in his hay.

~

Ten days later, Noah is walking the full length of our 120-foot barn. There've been two scares: once his left front leg buckled and he pitched forward, and once his back end gave out and he crashed to the ground. But if these moments fazed him, he didn't let on. Both times he simply got up unassisted, literally and figuratively shook himself off, and continued as if he understood that an occasional slip up was part of the recovery process.

Today, Norma Jean the turkey is all feathered bliss as she naps in the aisle. Noah lowers his head in a gentle greeting that Norma Jean knows she needn't fear. So, too, with Hannah the sheep, marching deliberately through the hay room as we walk past. She stops, lifts her head to Noah in confident greeting, and sheep and horse stand nose to nose for a moment. When she moves on, we do, too.

Noah seems to have gotten his sea legs—his steps are much more confident than they were ten days ago, though they're still a little cautious. He apparently knows that one could give out at any time. Noah delights us as he moves from one curiosity to the next. Sometimes it's a broom or a wheelbarrow filled with shavings, but far more often it's a living thing—a chicken or turkey, a sheep or pig.

Hazel the adolescent piglet trots up and lifts her pink snout in greeting. Many horses despise pigs; not this one. Noah lowers his head, and as the thousand-pound horse and fifty-pound pig greet each other, all soft breath and innocence, time stops. All is right in this world.

The return trip to Noah's stall is uneventful, except for the fact that the animal we weren't sure would live is pulling me quickly down the aisle. I know why, of course. We turn into his stall, and in an instant, his red head is once again eyeball-deep in the green hay.

~~~

A year after Noah and Cas were rescued, Marsha, the woman who'd wanted us to euthanize Noah, stopped in to say hello and to ask if we had room for another blind horse who was about to be seized in a cruelty case. It was just before feeding time. We hadn't spoken since I asked her not to call me back.

"Whatever happened to that horse?" she asked.

"Noah, you mean?" I responded.

"Yeah . . . Noah. Did he make it?"

I led Marsha out into large pasture behind the barn.

"Animals!" I called to the group grazing a couple hundred feet from us. "Bowie, Noah, Hazelnut! Crystal! Come on, guys! It's dinnertime!"

Five heads looked up, and Noah, the undisputed head of the herd, came galloping toward us. His head was high and he was running, pain-free, leading his friends to the barn for dinner. These days, in fact, Noah spends a great deal of time running—in part, because he's a Thoroughbred, and Thoroughbreds truly like to run, but I think also, in part, *because he can.*

Hazel's in Heat

Heat

It's the perfect word to describe a pig's ovulation. When a female pig ovulates, she is all raging hormones, all sex . . . all *heat*.

Petunia, our very first pig, was a whopping 250-pound potbelly. We all knew when she was in heat. With no male pigs around, Petunia would stalk male staff, walking right on their heels as they went through their day, then at the first opportunity insistently presenting her rear to them. She once trapped an unsuspecting electrician on his ladder when she refused to remove her rear end from the bottom rung. He, terrified, refused to step over her.

Nor is Petunia the only pig whose amorous activity crosses the line. Millie, the hundred-pound potbelly, insists on mounting Policeman, the old, gentle, thousand-pound farm pig, who wants nothing more on a sunny day than to plop his pink mass into the pile of wood shavings just outside the barn door and lie motionless, soaking up the warmth of the day. He's years (and a neutering) past thinking about sex. So Millie, unsatisfied, returns to her stall, where she plans her next assault on the next unwitting male as we wait for an opening in Hurley Veterinary Hospital's surgery schedule. Yes, for their health and our safety and sanity, we spay our pigs.

But now there's little Hazel's first heat.

Hazel, her mom, and three siblings were rescued last fall from a woman who had two hundred cats. And though she's but fifteen pounds and a few months old, she knows fully what she's after, and, in true pig fashion, is going after it with a vengeance.

She's mounting Policeman, then picking a fight through the fence with Piggerty, the female pig, foaming at the mouth and chomping her jaws in that "I'm going to *kick your ass*" way that's *so* uniquely porcine. A moment later, in a frenzy, she whips around and is mounting Winston, the black potbelly, and the two of them are a two-pig cha-cha line moving through the barn aisle, Hazel's front legs straddling Winston's rear end and her tiny back legs running to keep up as Winston, completely unfazed, goes about his business—searching for food.

Hazel mounts Hannah the sheep, lying in the aisle, who simply stands and walks away, turning to look at Hazel with a "What are you, crazy?" expression in her eyes. Norma Jean the turkey is diagonally across the aisle, pecking at treats on the hay room floor.

Hazel walks—*trots*, actually—toward the turkey. Out of nowhere Rambo the sheep appears, ever the guardian of all our fragile ones, blocking Hazel's best efforts to molest the gentle bird. He, too, is gentle but insistent, and a mere lowering of his head a few times, presenting his massive, curled horns, is enough to convince Hazel to look elsewhere. And she does. She continues down the aisle, intent upon satisfaction. We call the vet, hoping that a cancellation in his surgery schedule will mean quick relief for the hormonal Hazel . . . and the rest of us.

VegNews

VegNews. Think. Eat. Thrive. If it weren't for their Recipe Club, my Thursday night dinner would still be either vegan takeout from Mother Earth's, a local health food store with a fabulous deli counter, or broccoli and tofu in garlic sauce from King's Wok. By Thursday night, I'm often running on empty, and planning a meal is as appealing as major surgery on a vital organ. At least that used to be the case. Now, I have the *VegNews* Recipe Club, through which recipes magically appear in my inbox.

Click, print, shop, cook. It works for me.

And then there's the magazine. I have an entire winter ritual that revolves around reading *VegNews* from cover to cover. Weary at the end of the day, I draw the hottest bath my skin can tolerate, and only when the water has pulled the cold from my bones do I get up, put on my plush robe, pull up to the fireplace in the funky $20 Goodwill armchair that I've had since the '80s until I'm virtually sitting *in* the flames, pick up my mug of strong Bengal Spice tea, and spread *VegNews* on my lap. There's little in the way of fluff during these long, light-deprived months, so believe me: *VegNews* is a real treat.

But despite the fact that its devoted readership is a key target audience for us, CAS doesn't advertise in *VegNews*. With its soaring popularity and sleek, hip design come advertising rates that are simply out of reach. So imagine my shriek of delight when I learned that *VegNews* was offering $7,000 in free design and advertising to a non-profit whom they'd select through a simple application process. They'd pick the organization whom they thought best spread the good veg message; I knew Catskill Animal Sanctuary, with its 250-plus animal ambassadors, vegan cooking classes, and a national search for a vegan chef, would have a damned good shot.

The deadline was October 1. I put off completing the application, but not out of procrastination—I was simply way too "scheduled" in the weeks before the deadline: filling in at the barn for staff who were sick, injured, or out at conferences; making school presentations; taking trips to Manhattan; interviewing prospective chefs for our new full-time position. No matter. The application asked simply for our mission, our history/accomplishments, and how advertising in *VegNews* would help us expand our capacity. I'd add a few attachments—a recent newsletter, for instance—and be done in a couple hours.

Thursday morning, October 1. I head out to an early morning meeting with Sarah Chirhart, a local teacher interested in developing a relationship between CAS and her entire school. I've presented in her classroom numerous times; she understands and supports our mission and believes that her progressive new principal will be

The Great Dog Murphy

April delivers breakfast to the cows.

Caleb (L) and Patty (R). Caleb was rescued by a concerned neighbor; Patty was one of six animals CAS was delighted to rescue when Catskill Game Farm closed in 2006.

Puck, a tiny bantam rooster, weighs no more than three pounds but makes up for his size in attitude.

Claude the pig greets the day.

Ozzi with a young guest

Allen and Dozer. Like Dozer, many cows are extremely affectionate.

A few members of the Underfoot Family, our ever-changing group of animals free to roam the entire farm, graze in front of the horse paddock. Underfoot Family from L to R: Casey, Rambo, Hannah, and Aries.

Hope (L) and Echo (R) play gently. These two mares are among fifteen rescued from an irresponsible Saratoga breeder. Our farrier described the horses, who arrived thin and with wretched hooves and mangled manes, as "semi-feral."

Piggerty and Franklin run to the barn: it must be dinnertime!

A view of equine alley, the southernmost section of CAS

Animal Care Director Abbie Rogers with
Betty, a few days after her birth

Betty's brother Nash. In "Just Another Day at
CAS," I tell the story of their mom's rescue.

Visitors wait for a barn tour beneath a life-sized portrait of Samson the steer. The message "Peace to All Who Enter Here" is inscribed above his head.

"All of this food is cruelty-free!"
During our events, these signs are posted at our food tables.

Rambo and Hannah on early-morning patrol

Buddy

HEALTH CHECKS	
HORSE 1	4/1
HORSE 2 (Barni)	3/25
HORSE 3	
COW	3/28
GOAT	3/24
SHEEP	3/24
PIG	3/23
POT BELLY	3/23
TURKEYS	3/28
DUCK/GOOSE	3/31
Chickens	3/30-31
Rabbits	4/1

Buddy, who is blind, is featured on the jacket of my first book. He arrived at CAS with a dangerous panic disorder; his surrenderer did not tell us he'd been hit by a car.

Erasable white boards help all staff stay abreast of various daily and weekly routines.

Troy brings little Ozzi in for the night.

Jangles: now THAT'S a happy pig!

Rambo and his pal Barbie

Noah, a special friend

Gentle PeeWee, removed from a hoarder in 2004. Most of the woman's goats had already died of starvation.

Alex and tiny Lux. Our rabbits have come from a variety of distressing situations: dumpsters, fur farms shut down for cruelty.

My pal Franklin, back at CAS after his summer at Animal Camp

Lovely old Babe, who at fifteen is our senior steer

Sweet Norma Jean, rescued the day before
Thanksgiving

Farm manager Karen Wilson shares a moment
with Icy, who was found at a yard sale with a
"FREE" sign around her neck.

Doc (L), rescued in an Ohio cruelty case, and Malachi (R) share a meal.

My sweet Tucker, scratching an itch

Patrick and Phineas: *Isn't life grand?!*

Volunteer Julie Buono and I visit with Casey the horse, who lived in a junkyard and arrived at CAS thin, wormy, and covered in ticks.

Rambo: guardian, teacher, healer, friend

How I love him. Watching him age tears me apart.

Panoramic view of the center section of CAS

Little ones enjoy a peaceful evening.

Construction of our garden beds begins! CAS now boasts fourteen beds, with more to come.

Amos, rescued from the notorious Catskill Game Farm, grazes on a cool spring evening.

A portrait of Franklin as a piglet graces the entrance of one of our pig barns.

Now ten years old, Molly was rescued as a youngster from a hoarder in High Falls. Both she and Rambo were living with fifteen other animals, including Molly's dead mother, in one tiny stall.

open to a school-wide program centered around ethics. We're both energized and the ideas flow. When I leave, she says she'll set up a meeting with the principal and be in touch the following week.

I pull into the parking lot at Mother Earth's and hurry to the back to fill a quart container with one of their vegan soups. The sign above the huge crockpots advertises ADZUKI STEW, one of my favorites. I fill the ladle with the steamy stew of beans, eggplant, tomatoes, and spices.

"Kathy! Is that you, Kathy?" I hear behind me. The voice is scratchy, airless.

"Kathy," it resumes, and I turn. A woman in her sixties introduces herself. We'll call her Miriam.

"I'm a regular visitor to your lovely sanctuary," she says.

I'm surprised that she's not familiar. She pulls close to me. Her makeup is dense and precise: lip liner, heavy charcoal eyeliner, bright cheeks. Dark hair is pulled back tightly from her face.

"Kathy I've been looking for you *every time* I go to the Sanctuary and the last time I was there I believe *that pretty little one* oh what's her *name* isn't she named after some *season June or May* . . ."

"April," I offer.

"Of course, right, *April*, oh Kathy she's a *beautiful woman* isn't she and my gosh is she a worker! Anyway, I believe it was April who told me you were *away from the farm* for the summer working on your second book *your second book* that's so *marvelous* I don't know how you manage the time I mean *really I'd love to know* how you find the time to write when you're running *Catskill Animal Sanctuary* it's become such a *big operation* and oh all the building and painting you did this summer everything is so *new and gorgeous* it's really so impressive what you've accomplished your parents must be really proud *are you close to your parents Kathy?*"

"Well, my Dad and I . . ."

"Well Kathy I'm glad you're close to your father women should be close to their father *you know* you can't be a happy woman if you haven't been *close to your father* you know what I'm saying, Kathy, everyone needs to feel *loved and protected oh why am I saying this to you of all people you should know from love and protection* my god those animals are so content the last time I was there Rambo walked right up to me Kathy he was pawing my foot *what was he doing* pawing at my foot the tour guide said he wanted me to *rub his buttocks I thought I would die* oh Kathy he was a *handsome young man* too you know you've got very attractive *people* working at *the Sanctuary* Kathy and everyone is so nice . . ."

"Speaking of the Sanctuary, Miriam," I say, "I'm so sorry, but I've got to run. I've got an important deadline to meet."

"Oh of course you're so busy you must come here because you don't have time to cook *do you cook* Kathy you know it's really *important* for you to *take care of yourself* a lot of lives are *depending on you.*"

"Miriam, I've really got to . . ."

"Yes of course," Miriam charges ahead. "I just want to speak with you about my nephew Josh."

I just want to finish the *VegNews* application. Really. That's all.

"My nephew Josh is *to die for* Kathy oh he's dreamy it's really such a shame that he's a *flamboyant homosexual* you know the girls would really go for him he's tall and slender but you know he's always *gone the other way* I'm happy for him though he has a nice young man in his life right now he's black I believe or maybe a Latin actually I think he's a *black from Cuba* so *what does that make him I'm not sure* anyway Josh is a wonderful cook and I saw on the website that you're looking for a vegetarian chef and my Josh has *cooked from kindergarten* you know we should have known he'd *be a homosexual* the way he was *always in the kitchen . . .*"

"Is he a formally trained chef?" I ask.

"You know he'd be such an asset to Catskill *Animal* Sanctuary he's just so *flamboyant* always smiling and exclaiming and oh the gesturing with the *hands* he really is truly *to die for . . .*"

Miriam has barely taken a breath.

"Miriam!" I say, grasping her forearm. "I'm going to miss an important deadline if I don't get back to CAS. Josh can send me a résumé."

"Oh," comes her clipped response. "Can't he just come and *cook you a nice meal?*"

"It's a national search. He's welcome to submit a résumé. See you at the Sanctuary, I'm sure."

I hightail it to the cashier.

～⌒

There's no such thing as a five-minute check-in at the barn. A quick "How's the day going?" invariably turns into an hour of checking a building project, checking an injury, walking out with Abbie (our animal care director) to check a horse's weight, consulting with Karen (our farm manager) about multiple priorities, and acknowledging the good work of the volunteers. And then there are the free-range animals who walk up to say hello: Rambo the sheep, who comes up to request a butt massage though he's likely already received a half dozen; Chopper the portly potbelly, who throws himself down for a belly rub; Barbie, the lovely hen, who is always eager to be stroked, and so on. So today I plan to drive right past the big green building, walk straight into the house, and, as Miriam would say, *complete the VegNews application.*

But this is not to be. A pink tank is hauling ass through the parking lot and no human is running after it. It's Franklin, who has obviously found a way out of the newly constructed, nearly completed field we've added for him and his friend Tucker the cow. I dash into the barn. No one is there; no walkie-talkies are in their chargers. All are out with staff, who are out doing what they are supposed to be doing—working. If I drive the farm looking for someone to assist, Franklin will be in a neighbor's garden by the time we get back.

So instead, I race to get in front of Franklin before he can enter "Equine Alley," a long access road to four horse fields and our beloved motley crew of special-needs sheep and goats. Unfortunately, Franklin sees me and clearly remembers the drill from his adolescence, when he lived a blissful six months as a free-range pig and the game at the end of every day was to outrun whatever hapless human had been assigned to bring him back to the barn for the night. He sprints around me, grunting gleefully. It's a grunt that seems like a laugh, and I laugh right along with him as the two of us play chase.

～✦

Too much time later, with Franklin safely tucked away, I head toward my house. Fifty feet from safety, my assistant Julie rushes out of her office, jumps in front of my car, and puffs up her 110-pound body so that it looks a whole five ounces bigger.

"Okay . . . shaking in my boots here . . . what is it?" I say out my window, smiling at her.

"Call this woman right away," she says. "She's from the Pennsylvania SPCA that seized all those pigs last year. They won the case and are searching for help. She says they're desperate."

"Of course they are," I say to her. "They're trying to place *pigs*."

I call the woman, whose name is Callie. She confirms that they must place the pigs before winter.

"Reach out to the Pig Placement Network and to Tusk and Bristle," I say to Julie, referring to two wonderful organizations, "and see either of them can help."

I convene an emergency meeting in the barn: how quickly could we expand one of our pig pastures? Could we move the six horses in the pasture behind the barn to other fields, and how long would it take to put bottom rail around the entire pasture, which is essential to preventing pigs from rooting under the fence? Did Troy know if John the electrician was available to install radiant heaters? How much lumber was on hand, and was it oak, hickory, or pine, all of which have different strengths? Would additional pigs overwhelm Abbie?

It is 12:20 PM when I enter the house. I haven't eaten since 5 AM; my blood sugar is in my toes. Thankfully, my best pal Murphy is sleeping soundly on one of his many cushy beds; if I'm lucky, he'll continue sleeping until I complete the *VegNews* application.

I slap together an open-faced half-sandwich of avocado, salad greens, and heirloom tomato while I wait for the stew to heat. I'm ravenous. Just as I glance over at my beloved mutt, he's opening his eyes from his long morning nap. He seems both stunned and delighted that I'm standing just four feet from him.

"Hey, mutt!" I shout to him. "I love my doggie!" I say, getting down on my knees to wrestle with him. He's already snorting with glee; his tail thumps the bed quickly. After a good stretch, Murphy gets up and grabs a towel, brings it over and shoves into my thighs, his signal for a game called "Coming to Get You." Dutifully, I chase him around the house as he weaves through furniture, turning figure eights between kitchen, living room, and dining room. When I stop, he barks for more. I love that at thirteen years old my dog still has this vigor! At the moment, however, I wish his vigor had elsewhere to go.

I take the steamy mug of stew to my desk and pull up to the keyboard.

Name and address of organization:

Catskill Animal Sanctuary, 316 Old Stage Road, Saugerties, NY 12477

"This won't take more than an hour," I think to myself as my fingers fly over the keys.

Organization's Mission:

1. To rescue and rehabilitate twelve species of farmed animals
2. To raise public awareness of agribusiness and its implications for animals, humans, and the planet we share through innovative on- and off-site programming

The phone rings. I let it go to voicemail.

3. To serve the underserved by providing *pro bono* educational tours, school visits, and other such opportunities

History of Organization

Since we opened our permanent home in 2003, CAS has rescued over 1,600 animals from desperate situations: animals who had run out of options. Among them have been tiny Dino, a

250-pound pony who was the sole survivor of a Brooklyn arson;
Sammy the steer, who

Footsteps are coming through my hallway. I look up; it's Troy.

"Hi," he says.

"Hi," I say, and to myself, "Breathe."

"I'm sorry to interrupt, but there's a guy down here with a box of chickens he said he found on the side of the road; I'm not really sure what to tell him."

I hit "Save," then glance at Murphy, lying at my feet on his office bed. "Come on, dog," I say to him. "I need your help."

A man as thin as a whisper stands in the parking lot. A large box labeled LETTUCE rests on top of his black BMW. His khakis are crisp, his smile uneasy.

"I heard a rumor that there's more than lettuce in that box," I say, smiling.

Puzzled, he turns to look at the box, then back at me. "I'm afraid so," he says.

"What happened?"

"I was driving up 209 and saw this box very deliberately placed on the side of the road," he offered. "I figured someone pulled over for a flat and left the box by accident. I sure didn't expect this."

I looked in at "this": juvenile chickens, lots of them, their genders and breeds not yet discernible.

Over the past six years, CAS has taken in scores of discarded animals. We've taken a chicken found in a mailbox, chickens tied to a tree in Central Park, and a hundred or more animals found in dumpsters: ducklings, chicks, rabbits. And, of course, we took in precious Franklin, just weeks old, set aside by the pork farmer to starve to death. Yes, human beings discard animals as if they were garbage—many literally discard them *with* the garbage. But I don't believe this man's story. He won't look at me. My guess is that he's the same man who called two days ago to say that his chicken hobby "wasn't really working out" and he wanted to surrender his young chickens. Julie had offered to place the chickens on our waiting list; the man had hung up.

But all that is irrelevant now, because if we don't take these chickens, this man *will* leave them on the side of the road, or in a dumpster, or in the woods.

Before I lift the box from the hood of the truck, I look at the man squarely and say, "We have no room for these birds, so we're going to have to build another chicken house. If you'd like to make a donation, that would be wonderful." Then, making sure my voice is soft, I add, "Also, sir, I hope this experience will be an important lesson."

The man's face tightens and reddens. He reaches in his back pocket for his wallet, pulls out a thick stack of bills, and pulls two singles from it.

"No, you keep that," I say. I pick up my box of hungry chicks, turn, and walk toward the barn.

Abbie is with PeeWee, an ancient goat with an ulcer on her eye. PeeWee needs three kinds of antibiotic drops multiple times a day; she wears a lampshade around her neck so she can't rub her eye as it heals.

"Uh-oh," Abbie says.

"Yep," I say.

"How many?"

"This many," I say, lowering the box for her to peer in.

"Cute! So cute!" she exclaims. "Where are they going to go?"

"Gotta figure it out, Ab," I say to her. "I'll ask Caleb what he can build quickly, but can you figure out some temporary housing?"

"Sure thing," Abbie offers.

～〇

Some days are just like this. Not most, thank goodness, but honestly, way more than a few. Unexpected visitors, animal issues, and emergency rescues certainly do coalesce to derail a day. By the time I finally sit down again at my computer, it is 6:20. I look up at my blank desktop: no Firefox icon, no Microsoft Word icon. Nothing. No Excel, no Giftworks. Nada. No ability to access my document, no ability to start from scratch, no ability to send it if somehow the ability to type it returns.

No computer user is immune to bizarre glitches, kinks, and malfunctions. But this? *Now?*

I consider my options. Julie is gone. Having put in extra time earlier in the week, she's on her way to kickboxing. I can't open her computer without her password. My laptop is at David's house. There is no simple way to complete my *VegNews* application and e-mail it in.

At my feet, meanwhile, a patient dog sighs.

My thirteen-year-old dog. A free ad campaign in my favorite magazine—the magazine that speaks to *people who understand us.*

The decision takes a nanosecond. I hope it's one *VegNews* staffers could appreciate.

"Come on, mutt," I say, as I shut off my computer and rise from my chair. "Let's go for a walk."

Think. Eat. Thrive.

Pig Kissing 101

"So sorry about the delay," I say as I approach the twenty or so guests waiting patiently under the willow tree for the next tour to begin on this crisp Saturday in October. "We're missing a guide today." I tilt my head back for a good long drink of the water and juice mix I carry with me on tour days. It's going to be another day without lunch.

I take a seat on one of the benches, but just as I'm about to begin my introduction, young Franklin the pig looks up from his spot in the far corner of the pasture that borders our waiting area and I have an opportunity that must not be ignored.

"Kids!" I summon the group. "I need your help!"

I point to my friend with the pink skin and fuzzy ears, then say, "See that pig over there? His name is Franklin, and he loves children. Right now he's having fun digging in the dirt, but if you help me, I bet he'll come over."

"Do we have to go get him?" asks an outgoing little girl of six or seven.

"Nope. We just have to call him over. I'm going to count to three, then I want everyone to shout as loudly as you can, "*Franklin!*"

As his name leaves our collective lips, Franklin leaps the creek that divides his field in half and is trotting in our direction, grunting in anticipation. The group hovers around me.

"Whath he doing?" a wide-eyed child of five or six lisps through the hole in his top teeth, uncertain whether to laugh or to flee in terror from the 700-pound body barreling right at him.

I squat so that we are eye to eye. "What's your name, sweetie?" I ask of the little boy, who by now is nothing but breath and bulging brown eyes.

"Malcolm," he whispers, glancing furtively at his mom. Franklin is now a mere foot from us, pushing a soft snout into the wire mesh fence, his requests for company growing louder by the second.

"He's talking to us," I explain. "He's saying, 'Malcolm, come right here so I can meet you. I bet you'd be a great friend.'"

"Thath really what he'th thaying?" Malcolm asks.

"Abso*lutely!*"

Malcolm smiles.

"Hi, everybody," I say to the group. "I'm Kathy Stevens, founder of Catskill Animal Sanctuary. In just a moment you'll learn about the mission of CAS—who we rescue, how we make our choices, why we encourage all our guests not to eat animals like my friend here. But right now, we've got some pig kissing to do." (Sadly, as a result of the swine flu scare, this frequent ritual has been removed from our weekend tours. Many of the staff, though, still kiss the pigs, along with all the other critters.) A few chuckles emanate from the group, and one woman says, "I've been looking forward to kissing a pig all summer."

I edge over until I'm right in front of Franklin and offer my hand to Malcolm. He takes it.

"Pigs are very loud, Malcolm, and that's a scary thing if you're not used to it. But look: Franklin can't come any closer because he's behind this fence," I explain, touching the top rail of Franklin's pasture.

"Sit right here," I encourage him, and little Malcolm folds his legs and sits so that our knees are touching. "Hi, best pig in the world. Hi, you good, good pig," I say to my friend as I flatten my hand against the metal mesh so that he can push into it with his muddy snout the way he likes to do. "I love you, Franklin."

I take Malcolm's hand and hold it beneath mine, and watch the child's smile grow as Franklin greets him.

"He'th all muddy," Malcolm giggles.

"Yeah," I say. "Pigs need mud: It helps them stay cool since they don't sweat, and it helps prevent sunburn. Besides, you might be just as muddy as Franklin when you leave here!" I glance upward at mom.

"Not a problem," she responds, her smile as big as her son's. "This is worth a little mud." The group has gathered around us, and I sense another opportunity.

"Well, everyone," I turn around to address the group, focusing on the children. "I haven't given Franklin a kiss yet today, so I'll be right back."

I hoist myself up and over the fence, and step down beside my porcine pal. Franklin rubs his cheek against my thigh and oinks his most emotional hello. When I kneel, I am smothered in pig kisses: wet muddy snout against nose, cheek, head. I kiss him back, then smile to the group. Most are laughing with delight; one woman looks like she wants to grab her child and flee from what is surely a demonic cult. ("Carlton, *they actually kiss pigs*," I imagine her saying to her husband over their pork chop dinner.)

"Anybody else want to kiss a pig?"

Before she either faints or vomits, the pork chop eater does, indeed, take her child and head toward the parking lot. In the meantime, two young girls are squealing with glee, entreating their parents.

One at a time, Dad passes each of the girls over the fence. Franklin, of course, is beside himself, and the girls are instantly both filthy and in love. "I love you, Franklin," the older one says. "I love you, Franklin," the younger one mimics through delighted giggles as a cool snout greets her.

I pass the human packages back over the fence.

Malcolm, frozen in place on the other side, looks up at me, his eyes saying everything.

"Ok, trooper," I smile as I hold out my hands to help him over. "Ready to have some fun?" He looks at his mother, who asks simply, "What do you think, Malcolm?"

Malcolm utters something inaudible as he stands and raises his hands to be lifted over the fence. Mom passes him to me. At first, Franklin's mass, presence, and proximity overwhelm him. He stands frozen, his back to the forward pig and his face buried in my shirt.

Franklin pokes the small of Malcolm's back, and a muffled giggle emerges. Franklin pokes some more: Malcolm's back, his fanny, his legs. The little boy's giggle becomes a laugh, his laugh a guffaw as Franklin ramps up his enthusiasm.

"Franklin, you're tickling me!" Malcolm shrieks. By now he has turned, and Franklin is laughing deliriously with him, that open-mouthed pig laugh that, once witnessed, is not soon forgotten.

"You didn't *tell me that pigth were ticklerth,*" Malcolm cried.

"I know I didn't, Malcolm! Franklin wanted to surprise you!"

I glance up at Malcolm's mom. A smile as wide as the Mississippi is spread across her face.

It's a
Hot Day Here

I don't see a single leaf waving. It's probably ninety-five degrees today; the air is thick and still as death. But I don't turn on my air conditioner, for quite simply, it uses too much electricity. Global warming. The earth is pleading for our help, so I'm making concessions where I can. Like turning on the hot water heater only thirty minutes before my shower. Like pooling my errands so I drive less. Like deriving 100 percent of the Sanctuary's energy from a newly installed solar system. And yes, like going without air conditioning when I want it. Being vegan, of course, is the biggest thing I do. Lessen the footprint. I'm trying.

The animals, after all, don't have air conditioning. They have shade, sure, but on days like this they'd love to be standing right in front of that machine that magically takes hot air and turns it frigid.

Today, we all sweat. I sweat at my computer, shedding first my shorts and then my shirt as I hope against hope that no stranger appears in my doorway (as they often do). *Quelle surprise!* I hear myself rationalizing why a "director" is working in her underwear. Julie bakes in her office, refusing, bless her conscientious heart, to turn on her air conditioner. April, Abbie, and Alex look like they've been playing under a sprinkler—and no doubt wish they had.

Among the animals, the pigs fare best. They loll in the mud baths we make for them under the willows, find the shade trees and dig up the cool earth, and grunt with glee when, as they leave their pasture to come to the barn for nighttime, Abbie sprays them with cool water from the hose.

The sheep, recently shorn but nonetheless trapped inside their dense coats, pant heavily. We wonder if we'll need to hose them down, too. The horses and cows find the shade and stay still, but two elderly cows are nonetheless dangerously close to heatstroke.

Yet it's the smallest of all, the poor broilers, who struggle the most. They've got the coolest spot on the farm—their house sits under a canopy of massive willow trees, and their yard is literally a foot from the pond. Still, as the thermometer inches toward a hundred degrees, every single broiler is holding his wings away from his body and panting heavily. We mist them until their skin is wet and hold them in shallow troughs of cool water, hoping their body temperatures will drop.

Chickens, you see, aren't supposed to weigh fifteen pounds. But those that miraculously escape slaughter *do* weigh that much, often more. Agribusiness has created "Frankenbirds"—chickens that grow at freakishly fast rates (greater profits for the producer) yet have such high death rates that agribusiness itself has created the term "flip-over syndrome" because it finds so many young chickens lying on their backs, feet pointed skyward, dead from violent heart attacks because their hearts simply can't take the rapid growth. A number of our own birds have died in this way.

Yet we humans continue to eat them. Tortured birds, caged pigs, terrorized cows. Under agribusiness, the tiniest concessions to their well-being are long gone. These animals are commodities, period. It doesn't matter that they suffer mightily. It doesn't matter that they are so very much like us, or that pain is pain and suffering is suffering whether it is inflicted to a human or a dog or a chicken. It feels the same, doesn't it, no matter what one's species?

So just like the 300-pound human is having a harder time on this hot humid day than I am, so are our chickens. They're desperately gasping for air.

Home for the Holidays

Thoughts on Easter Sunday

In the midst of such grandeur, there is the wretchedness we have wrought.

David and I have just returned home from vacationing in Utah's Zion National Park. It was a glorious week in God's country, with the exception of one moment that will be seared into my memory forever. It haunts me now as I walk the grounds of Catskill Animal Sanctuary on this Easter Sunday.

As we set out on the first of what would be epic daily hikes, David and I spotted a small pen filled with horses and mules tacked up for the day's trail ride. While I am *far* from being a fan of these "rent a horse" outfits (the animals lead a life of drudgery, at best, and when they've outlived the purpose that humans have ascribed to them, most go to slaughter), these animals actually looked terrific: their weights were ideal, their hooves were in great shape, their coats were healthy. I was also heartened to learn from the guy running the outfit that the mules and horses work for two days and then have a day off. Many "dude ranches," trail rides, and carriage horse operations give their "employees" (i.e. the horses) only one day off per week.

"They're luckier than most," I said to David.

And so we headed out, exploring the glorious Zion Canyon, spending an absolutely delightful afternoon, until we got to the fork in the path.

"Let's go this way," David suggested.

After a few hundred steps, to our surprise, we arrived at the barn where the horses and mules lived, and my fantasy of spacious pasture and ample shelter in which the hardworking animals would enjoy their day off was instantly shattered.

Over forty mules and horses were packed into a turnout pen that was, at most, sixty feet square. On their day off, these animals, mostly mule geldings, had only a tiny, dusty, shadeless pen. No trees, no grass, no room to frolic. The single run-in shed was way too small to accommodate even a third of the animals, which presumably means that when the scorching sun is high in July and August and the thermometer reads well above a hundred, many animals must simply endure it.

I checked their water. It was green. Thick algae lined the entire trough and an oily scum floated on top. Shit. People call Catskill Animal Sanctuary when they see animals enduring conditions such as these. *Can you do anything?* they ask, naively hoping that we'll either take the animals, file criminal charges, or guide them through the process of doing so.

"No," I say . . . more often than I would like. Like anti-cruelty statutes throughout the country, New York State laws are far too lax and far too vague. Though some counties are far better than most, in general, law enforcement is resistant to prosecuting all but the most horrific cases. If an animal's weight is good and water (even if it's filthy and not potable) and shelter (even if it's a tree) are available, no law is being broken. All one can do is look for ways to engage the owner about his animal husbandry practices and monitor the situation to ensure it doesn't get worse. It often does. Then we *can* act.

I sat cross-legged in the dirt outside the paddock. Four animals—a mule and three horses—approached eagerly. Three were old mares, their huge bellies sagging, their bodies scarred from the kicks and bites of more dominant animals. These animals were far too old and worn out to be ridden. I know their fate. Like virtually all horses who outlive their usefulness to humans, these will likely be slaughtered.

They approached eagerly. "Hi, girls," I said. "I'm sorry I don't have treats." Still, one nuzzled my face, and another lowered her head and allowed me to massage her cheeks.

Twenty days a month, year after year, these animals work hard, carrying mostly those who know nothing about riding—nothing about sharing an experience with an animal—back and forth along the same dreary path, in temperatures that rise to 110 degrees. Their reward for their service? A few days off in a hot, tiny, dusty pen drinking contaminated water.

Across the United States, there are tens of thousands of such animals . . . as well as those in zoos, in traveling circuses and rodeos, in canned hunts, at racetracks, and at theme parks. They endure their respective wretchedness because humans want to be entertained by them. But would our lives diminish qualitatively if we no longer attended a rodeo, a dog or horse race, or a circus? Would our vacation be less enjoyable if, instead of riding a bored and overused animal, we hiked the trails ourselves? If instead of taking the kids to SeaWorld to watch dolphins jump through hoops, we took them to the *real sea* to swim and build sandcastles?

That human beings feel entitled to use animals for any purpose we determine to be fun or profitable is a level of callousness or obtuseness or disregard common to most of us in the developed world. For most, it's an unquestioned assumption of privilege; for a few, it may be a resigned sense that "it's just the way the world is." It never occurs to most of us to question the status quo.

I'm asking you to do so. And I'm asking you to acknowledge your own role in the suffering of animals.

It's this very concept that Matthew Sculley discusses in his book *Dominion*. It's not a new book, but in its unflinching presentation of what animals endure at our hands and its call for mercy, for my money it's one of the best. Sculley asks whether "man's dominion over animals," as discussed in the Bible, suggests dominance or caretaking. Providing snapshots of the brutality and misery inherent in many animal industries, Sculley makes the obvious case that humans do, indeed, dominate all other animal species with breathtaking disregard of their suffering. Very few of us consider these issues. When we do, in my view, there's no argument to be made. We have enslaved them, plain and simple.

But this domination/subjugation/oppression model was not God's intention, Sculley argues. No, by giving us "dominion over animals," God intended us to be their caretakers.

Today is Easter Sunday. As I walk around this precious animal sanctuary on this symbolic day, joyful animals, allowed to be themselves, surround me. Policeman, a thousand-pound pig rescued from a Bronx apartment, is one happy camper stretched out on the sunny side of the shavings pile. Molly the cow and Sammy the steer playfully butt heads. I smile as Helen the blind calf licks the face of Andy, the young gelding still too weak to be turned out with other horses. No matter: Helen takes such good care of him. The goats play "King of the Mountain" at their rock pile, and, in one of the surest signs of the coming of spring, the hens are busily scratching in the dirt for whatever edible tidbit they can uncover. At the far end of the farm, Buddy the blind horse rolls blissfully on the cool ground. I watch these lucky few, and I remember the Zion horses.

Jesus suffered mightily. So do the animals. The power to change this is in our hands.

Happy Easter.

We Ate Pumpkin Pie

I spoke on Thanksgiving afternoon at the Berkshire Vegetarian Society's "Living Thanksgiving," held at the United Methodist Church in Lenox, Massachusetts. Roughly one hundred people were there; no turkeys died to feed us. Rather, there was table after table of everything *but* dead animals: cranberry sauces and pilafs, potatoes and yams, and entrees like loafs and soufflés and tortes and stuffed vegetables. And then, of course, there was stuffing . . . lots and lots of stuffing. Isn't Thanksgiving dinner really about the stuffing, after all?

My beloved pal Murphy the dog was invited in, and he had a grand time in the midst of so many good smells, so many animal lovers. Murphy begs very politely, and more than one guest succumbed to his patient entreaties.

I read from *Where the Blind Horse Sings.* I talked about the life-altering lessons learned from broken animals made whole again. Questions from the audience were wonderful and provocative, and in the end I invited the group to visit Catskill Animal Sanctuary the following day. It was a lovely afternoon.

I pulled up to my house at 6:30 under a nearly full moon. I walked up the back steps, stripped off two jackets, and walked immediately to the oven with the pumpkin pie given to me as I left the church. I piled old newspaper, then cedar kindling, then locust and oak logs in the fireplace, and an amber glow lit the living room.

Just outside my front door, I heard the horses. I peered out to see old man Maxx and his friends Callie and Hazelnut. Their heads leaned over the deck railing; their ears pricked forward in eager anticipation of a friendly greeting and a treat. No carrots to be found, however, and my fruit bowl, often piled high, was empty.

I wondered. . . .

I pulled the pie out of the oven, shoved one arm then two into my red corduroy jacket, then my green one, and walked out to the deck.

"Animals," I whispered, "animals . . . *look what I have!*"

Murphy got the first bite, shoving his snout right into the center of the pie. Callie was next. She sniffed tentatively: what kind of treat is *this*? and then licked the surface, ever the lady. Hazelnut did the same.

And then came Maxx. Maxx, the thirty-something-year-old gelding surrendered to CAS after his owner died of cancer. Maxx, the gelding with his harem of six mares. Maxx, the pumpkin pie lover who took one perfunctory sniff and then smashed his muzzle so forcefully into the pie that I nearly dropped it, and then again, delighting in its texture and sweetness.

"*Happy Thanksgiving, Maxx!*" I exclaimed, laughing heartily.

Murphy and I stayed on the deck for another few minutes, surrounded by horses content to remain right there with us, drenched in moonlight, soaking up the love, and savoring pumpkin pie.

Merry Christmas, World!

I could be with David in Hawaii, my Dad in Florida, my brother in Virginia, my sister and her wonderful brood in Michigan, or my grandmother, aunts, uncles, cousins—the whole maternal clan—in Nashville.

Instead, clad in long johns, jeans, boots, gloves, hat, T-shirt, turtleneck, fleece vest, and jacket, I'm scooping poop at Catskill Animal Sanctuary, assisted by the great dog Murphy, and I couldn't be happier.

With our two animal caretakers either on vacation or taking the day off, I'm in the barn—per usual—on Christmas Day. WAMC, the public radio station, is airing Christmas essays, including David Sedaris's hilarious account of his single day working as an elf in a shopping mall. April and Allen and Alex are here with me. Quickly and effortlessly, we divide up the morning feed routine: April and Allen feed the "outside" animals, mostly big animals in big pastures the farthest from the barn; Alex feeds the "barnyard" animals, the rabbits, ducks, and chickens in seven different shelters clustered closer to the main barn; and I feed the menagerie inside the barn: eight special needs horses whose age or condition have earned them a permanent spot there; the eighteen potbellies and big pigs who appreciate the heated stalls; twelve goats; Lama and Jack, our two blind (or nearly) sheep; an eclectic assortment of birds—five broiler roosters, Norma Jean the turkey, roosters Sumo, Rocky, Doodles, and Scribble . . . and so on. Today, a few extra treats are placed in each feed dish. Today, every single animal gets a kiss. Every chicken gets held, every pig is massaged, every horse muzzle has a kiss planted on its smooth, warm center.

"Umh, umphhh," Franklin the pig grunts in gratitude. And Norma Jean, our rescued turkey, settles into my lap—uncertainly at first, but with each new breath, she lets go a little until her eyes are heavy and she's asleep.

I steal away mid-morning and an hour later return with three-dozen pancakes. Christmas brunch in the barn! We pass juice and maple syrup, and vegan dietician George Eisman and his girlfriend, Melanie Carpenter, come by with one of Melanie's extraordinary desserts. So what if we've just finished a pound of strawberry pancakes apiece? It's Christmas! We dive into Melanie's chocolate mousse pie. This food is all made without animal products. And it's all divine.

Outside the kitchen door, Franklin grunts. "Can I come in?" he pleads. We're tempted, but as you know, Franklin is no longer the five-pound piglet who arrived at Catskill Animal Sanctuary three winters ago. He is 700 pounds, and a 700-pound pig loose in a kitchen wouldn't be pretty . . . not even on Christmas.

I grab a handful of pancakes and slip out the back door. "Merry Christmas, best pig in the world," I whisper to my friend, who gleefully gobbles the pancakes. "Come on, boy, it's time to go back to work," I say to him, and Murphy, Franklin, and I head down the drive to clean the goose house.

Merry Christmas, World.

When Winter Kicks Your Ass

When snow comes early in the season and hangs around for weeks, freezing into hard chunks that could slice a chapped finger, winter kicks your ass.

When a second (and then a third) snow accumulates on top of the first, winter kicks your ass.

When Alex has to use a shovel to get to the snow blower and the snow blower to get to the plow truck and the plow truck to clear the parking lot, barn doors, and roads to the outer pastures—all this before feeding can even begin—winter kicks your ass.

When the wind is so bitter that not even the hardiest cows want to leave their cozy barn, winter kicks your ass. (It's your job, after all, to keep that barn clean for your friends, and eighteen cows produce a lot of poop.)

When everyone in the barn is swearing like truckers because they're down to their last set of toe warmers, winter kicks your ass.

When the animals are cozy in their heated stalls but the heater in the only warm room for humans shuts down on a holiday when no one will repair it, winter kicks your ass.

When Jangles the pig wants to do nothing other than snuggle under the hay in his heated stall, but you know he must get up and move those stiff joints, and when in frustration, he whips around a little too far and suddenly pig head and human head collide for one painful, dizzying moment, winter kicks your ass.

When all but your most dedicated volunteers disappear because, really, this weather is just too much, winter kicks your ass.

When the wind is so strong that the simple act of moving forward is a Herculean one and your job is to deliver forty bales of hay through that wind to the horses and cows in the outer fields, winter kicks your ass.

When what looks like driveway is really black ice and your feet fly from under you and you crash to your elbows, and the grain buckets you were carrying are suddenly flying like Frisbees, winter kicks your ass.

When the feed delivery truck slides on a patch of ice into the ditch followed by the tractor that was meant to retrieve it, winter kicks your ass.

When on that very same day the tractor trailer delivering eight hundred bales of hay slides sideways into the ditch, blocking the exit for cold and cranky staff and volunteers, winter kicks your ass.

When four people in a single day are desperate to place their animals because their homes and farms are in foreclosure, and when even though the *last* thing you want to do in these grueling conditions is take in more needy animals, you consider it because, quite simply, that is your mission, winter kicks your ass.

One Perfect Day

My dog Murphy truly is my best friend. As far as love goes, I have known none greater. He is my teacher, my child, and by his choice, my constant companion. He is thirteen. One day, we will take our final walk in the woods, and I will need to remember this day.

Murphy woke both Spot the cat and me at five this morning. Spot, formerly feral, moved into David's house a few weeks ago after a couple months of progressively bold overtures: first, she stared down at us from the cliff just beyond the porch for a few minutes, then vanished into the woods. Then, for a couple weeks, she rested under a porch bench as long as David was alone. As soon as Murphy and I pulled in, however, she ran for her life. And so it went. After a while, we were feeding her just outside the house, then just inside the door . . . you get the picture. Now, Miss Formerly Feral sleeps either curled up with Murphy on his bed or between David and me in ours. So much for the "once feral, always feral" notion.

Anyway, this morning, the three of us ventured down David's twenty-four stairs for breakfast. Murphy's: chicken, kale, pumpkin, and a tablet each of Vitamin C, Vitamin E, fish oil, and Bufferin. Spot's: Tiki Cat's "tuna on rice with prawns." Coffee for me; I would wait until the sun came up for my smoothie.

(Yes, my animals eat meat. A few vegan friends have taught their cats to be mostly vegan; one friend's cats eat chickpeas and alfalfa sprouts. We've all done our homework; we've all come to our own conclusions. I believe cats are true carnivores; dogs, more omnivorous. However, having read the work of Richard Pitcairn, Gary Null, and others, I also believe that meat is the healthiest protein for my dog. Some see this practice as inconsistent with my beliefs. I don't. I'm a human who benefits on every level from a vegan diet. He's a dog who, in my view, benefits from a clean diet—absolutely no processed food—that replicates the one his wild relatives have consumed throughout time.)

Effortlessly, the mutt preceded me back up David's stairs, just as he had done the night before. The day will arrive when he needs the "dogwaiter" David is building to hoist him from bottom floor to top. But for now, each easy climb feels like a rejection of aging, and I celebrate every single one. Before I said, "Up, up!" Murphy jumped up, placing his front paws on the bed and waiting for me to lift him. He circled a few times, found his spot. "I love you, muttster," I said as I climbed in bed after him. I said it loudly; I had to if I wanted him to hear me.

I propped three pillows against the headboard, and sat upright, legs crossed. I closed my eyes and felt my breath deepen and slow. "Thank you," I said as I began my morning ritual of gratitude, affirmation, and visualization, ignoring (or trying to) Spot's motor revving right next to me. I spent a long time on Murphy, feeling grateful for his health, affirming his youthful vigor, sending him love and healing energy. Mind you, I have little or no formal training in any of this stuff. But I have read a great deal; far more importantly, I have always instinctively believed in the power of belief. Beginning my day consciously saying thank you for the extraordinary beauty in my life, sending good energy to those I love, and visualizing things I want to manifest is important to me. And maybe, *just maybe*, doing a little of this work for my mutt will contribute to his ongoing health.

I feel the same way about the power of touch, so I spent a few minutes massaging his shoulders, his hips. Murphy stretched out his rear legs and yawned his release.

Back downstairs, Murphy and I positioned ourselves on his jumbo bed just in front of the woodstove. He gnawed on a fleece shirt pulled from the laundry basket; I struggled to write. My brain felt like a bog. Murphy stood at the door and barked once, sharply.

"Okay, mutt, thank you for telling me," I yelled, and dragged his bed to David's porch.

I returned to the sofa and my laptop. I watched Murphy atop his perch, lifting his nose to catch the scent of deer, bears, foxes, Canada geese. Once or twice he barked; twice he sprinted across the meadow after a deer, a pheasant, or a phantom.

"Good, guard the house, boy!" I bellowed out to him, hoping he could hear me as he wandered back to the porch.

"*Woof*!" he said as he hopped back on the porch.

Great. The single sharp bark that means "Come here right now, Mom!" I'm supposed to be working, after all.

"What?" I asked him.

Murphy walked to a large planter that holds everything but plants: umbrella, clothespins, a dozen or so tennis balls, and an assortment of Murphy towels. He grabbed a faded purple one and backed up, taunting me into one of many games we've played throughout his life.

"*I'm coming to get you!*" I yelled, and as the yellow mutt hopped off the porch and trotted through the yard, I chased him. He turned, towel wadded in his mouth. Time, evidently, for his second game.

"*Give me that towel!*" I bellowed as I grabbed it, pulling as he did—classic tug of war—but pulling more gently than I used to, and releasing more gently so that he didn't crash to the ground as he pulled with all the strength a ninety-year-old man could muster. I got on all fours, looking him straight in the eye as I held the towel with one hand and Murphy dragged me around the yard. When his energy was spent, he lay down, gnawing on his treasure until he was asleep in the sun.

I plodded through a chapter as though it were sludge. Each sentence was a struggle: a rarity for me, thank God. How grateful I was for Murphy's frequent interruptions. His cold nose on the glass door, telling me he wanted to come in. His frequent rising from rest, looking me straight in the eye, and proceeding to the refrigerator, wanting its contents, getting what was allowed. Another mid-morning game, another request to be let out to "guard the house." It had been a few days since Murphy had been fully "himself," the high-energy, exceptionally interactive dog who follows me from room to room (even lying outside the shower door). With his umpteenth request for mama time today, I finally understood why I could not write: *I want to be with my dog.* It is very simple—he is thirteen. I do not know how many more days like this he will have. I hit the "Save" button, shut down the computer, and load the mutt and his mattress into the car. December deadline be damned.

～○

In October 2008, I rushed a twelve-year-old Murphy to Kingston's Animal Emergency Clinic after he nearly collapsed outside the house. X-rays revealed a large mass on his spleen, and blood work suggested that the mass had ruptured. My doggie was bleeding internally. Other than euthanasia or allowing him to bleed to death, an emergency splenectomy was the only option.

The vet explained that Murphy had hemangiosarcoma, a "very aggressive" form of cancer, and that, if lucky, he had two months to live. But there were three reasons why I instructed her to perform the surgery. First, I was certainly not prepared to say goodbye to my best friend right then and there. Secondly, though the chances were three out of four that Murphy had x, there was a one in four chance that Murphy had y: benign nodular hyperplasia, the operative word being "benign." And finally, though my best buddy likely had a rapidly spreading cancer that would take his life in thirty to sixty days, maybe we could buy him more time. My friend Kris Carr, diagnosed with Stage 4 cancer in 2003, is still going strong seven years later, inspiring others around the world to take care of their health and their lives through diet, meditation, positive thinking, and community. I do believe that love, touch, positive thinking, and diet all have transformative power. While my beloved dog was recovering from surgery, I would learn what I could do to improve his chances to surpass his terminal diagnosis. Kris and thousands of others have done it. Maybe with my help Murphy could, too.

Well, Murphy was that one in four. His tumor was benign, his life since that eventful night mostly wonderful. Along the way there've been important lessons. Far faster than his joints or his vision, what's aging in Murphy is his digestive system. We learned through two harrowing incidents that Murphy can no longer tolerate processed food, and I learned not to leave *anything* on the countertop. His age notwithstanding, the mutt is still a master thief who will happily inhale an entire loaf of bread, jar of peanut butter, or tub of Earth Balance. Once in a while, generally when we've overdone it the previous day, Murphy has low-energy days when he wants to do little other than sleep, eat, mosey to the barn to sniff around for something on his "Do Not Eat" list, and bug me once or twice to chase him around the house when he invariably rallies in the late afternoon. On days like this, I try hard to simply enjoy watching him sleep as I will his energy to return. Thankfully, he has far more days like today.

~⌒

"Ready to go for a ride, mutt?" I turn and shout at him as he positions himself squarely in the middle of his plush bed. My Subaru wagon is little more than a Murphmobile—in fact, I bought it for him, knowing that being hoisted into my Ford pickup would, over time, become difficult for his aging joints. Now, the back seat stays down permanently and the entire back of the car belongs to the boy. I pull his bed up until it touches the front seats, so that often his head is just inches from mine as we tool around town.

We head to the Sanctuary. It's the middle of hunting season; the wooded portion of our property is thoroughly posted with NO HUNTING signs, as are the hundreds of adjoining acres. This time of year, it's one of the few places where we can safely walk.

Miles of beautiful trails meander through this land. As we enter the woods, the first trail, a large loop, is relatively easy and only about a mile long. I'll gauge the mutt's energy to determine whether we'll do the complete loop.

"Ready to go for a walk, Murph?" I shout to him, but am barely able to grab him before he leaps from the car and jars his old joints. No matter. Despite having been gently deposited on the ground, he takes off in hot pursuit of his favorite rodent; he hasn't a chance in hell as the squirrel leaps through the air and in a nanosecond is scolding Murphy from a branch fifteen feet above us.

"You *asshole!*" the squirrel yells. But Murphy is oblivious. His nose is down and he's trotting in front of me, knowing from thirteen years' experience that when dead, crunchy leaves are ankle deep on the ground, furry friends are plentiful. Without warning, Murphy takes off. He's picked up a deer scent and is scrambling up the cliff toward the trail that runs along its ridge.

"Murphy, stop!" I yell to him, as he continues up the cliff's steep face. My command is as effective as it's always been—even more so now that there's an excellent chance he doesn't even hear it. Yes, although it's not exactly a sprint, my thirteen-year-old dog *is running up a steep hill after deer.* He is driven by DNA, utterly overcome by excitement, and his body is up to the challenge. Nose to the ground, he turns on a dime, running now across the cliff, not up. In heavy, awkward muck boots, I can't quite keep up, but as I watch my boy go, I am bursting with joy.

We are back on the trail now, perhaps a mile from where he picked up the scent. Fortunately, an enormous downed tree slows his momentum. "Good, get the deer!" I praise him as I catch up to him. He is panting hard. He wants to continue, of course, and he feels so good that for a moment I consider it. But I figure that over the next few days he'll feel what he's already done. Reluctantly, I turn toward home.

Murphy doesn't budge. "Why, Mom?" he's asking. "Why can't we keep going?"

Back at David's house, Murphy's energy remains high. We take a couple short strolls to look for deer. After dinner, he drags one towel, then two, finally three to the yard to gnaw himself delirious. Eventually, we climb the stairs to go to bed. Spot, who has kept it warm, is delighted to see us. I bend down for our ritual head butting session, while beside the bed, Murphy circles a few times, plops down on his mattress, and sighs. It's been a long, good day.

"I love you, dog," I whisper as I lay down next to him to give him a good rub down.

"I know you love me, Mama," Murphy answers back with a slow blink and enough kisses to make my skin hurt. "You tell me a hundred times a day."

"I know I do, but I need to tell you a hundred and one," I say.

"Okay, Mama," the mutt says back to me. "I love you, too."

On
Anthropomorphism

An ancient, tiny pony named Dino was the first animal to call CAS home. He'd been the sole survivor at Brooklyn's Bergen Beach Stables after a teenager threw a match into the hay room. Twenty-three horses succumbed, but Dino, probably the oldest and undoubtedly the smallest, had survived. Dino had kicked, and kicked some more, until his door came crashing down, at which point firefighters rushed in and dragged his burning body to safety. The next day, the New York *Daily News*, a well-known New York City tabloid, ran a front-page photo of a firefighter sharing his oxygen with the pony. Dino was one brave little man.

Dino lived with us for six and a half years; we lost him in August 2007 when he was forty or more years old. Immediately, we contacted those who'd been closest to him over the years and sent an e-mail out to our volunteers. But because he'd been so beloved, I also wrote an announcement of his death and sent it to the *Saugerties Times*, our town's weekly newspaper, and to the New York *Daily News* as well.

We were overwhelmed by the outpouring of grief, love, and support. Letters, cards, e-mails, and gifts arrived from around the country. People brought flowers, carrots and apples, mementos. We set up an altar in his stall: simple hay bales stacked on top of each other, on top of which we put candles (contained in votives, and placed on small plates for extra protection), photos, and flowers. More gifts arrived. We held a simple memorial service, and in the end, scattered Dino's ashes around the farm. I was delighted when Will Dendis, editor of the *Saugerties Times*, called for an interview, and more delighted when his piece "Requiem for an Equine" appeared in the paper that week.

Yet while the article was written with sensitivity, how disappointed I was when I got to the final paragraph. It read:

"Perhaps Dino's story as related by Stevens might sound a tad anthropomorphic, a case of an animal lover projecting human qualities onto the dramatic life of a very old pony. But the strong reaction and pure volume of outpouring over his death means at least one thing: this little pony had a lot of friends."

I had told Dendis that Dino was courageous, and that he had an indomitable will to live. He'd drawn his conclusion from those two descriptions. But how else would one describe the sole survivor of an arson to which twenty-three horses larger and younger than Dino succumbed? How else would one describe an ancient, severely arthritic, partially blind pony with extremely limited lung capacity and a throat so filled with scar tissue that *swallowing* was difficult, who nonetheless greeted each new day with enthusiasm? When one has seen animals with fewer health issues give up and allow death to come, what other words would describe a pony who, despite all these challenges, whinnied when the first of us entered the barn each morning, and stuck his head out into the aisle to say good morning?

Not only do I dislike the term anthropomorphism, but I don't buy it. I don't think that when we talk about a dog being worried or a chicken being excited or horse being depressed or a pig being jealous we're ascribing human emotions to animals. I believe the dog is worried, the chicken is excited, the horse is depressed, and the pig is jealous. In my view, the term anthropomorphism is used either malevolently, for instance, by scientists trying to assuage their guilt or deny their humanity as they justify horrific experiments performed on animals, or mistakenly, by people who know little about animals and thus accept the popular notion that they are fundamentally far more emotionally limited than humans.

At Catskill Animal Sanctuary, we work with animals every single day. While they arrive here broken and fearful, over time we watch them blossom, often into enormous and unforgettable characters. Interestingly, the process is similar from animal to animal, and no doubt mirrors the experience of human beings recovering from a history of trauma, violence, or neglect. First they trust their caretakers—those who give them food, shelter, and love day after day. Then we watch in delight as they generalize to visitors. Indeed, there is no greater joy than participating in the transformation of these broken spirits and watching them evolve as dark memories are replaced by consistent positive experience. And once they know in their bones that *only good happens here*—that's when the magic really begins.

Ask anyone at Catskill Animal Sanctuary or come visit or volunteer and discover for yourself: we'd be hard-pressed to name an emotion that animals *don't* possess. They display love, tenderness, joy, curiosity, impatience, anger, jealousy, grief, and a host of other emotions generally considered the exclusive domain of humans. The greatest among the animals display things like compassion and courage. Just ask any of us about Rambo's empathy (his lessons merited four full chapters of my first book) or Noah's compassion. Conversely, just ask any of us who's the most selfish horse on the property. It's unanimous: Big Ted.

Franklin the pig has a delightful sense of humor. Rambo the sheep is wiser than any human I know. Before he died, an old steer named Samson licked my face over and over—until he took his final breath. A dozen people witnessed this. I believe this old man, whom we'd rescued from a hellish situation, was saying goodbye. I believe he was saying thank you, and I believe he was telling me he loved me. If you still need to call me anthropomorphic, go ahead. I'll pity you for the richness that you're missing. More importantly, I'll pity the animals, who suffer mightily because we humans fail to see them for who they are.

I am delighted to be among these animals, blessed to have them as my teachers, and obliged to share with the world what I have learned. *If only we all knew this,* I say to myself. If all of us knew in our bones that animals shared the same feelings we do, would it make a difference?

If I'd Had a Camera

I wish I'd had a camera. But then I'd have ruined the moment. I've never once walked in the barn during work hours and not seen a human. Not in seven years. But today it happened. Alex was up in the large hill pasture reinforcing fence. (Two mules arrive Sunday—their family's farm is in foreclosure—and one of them, Blackjack, is nicknamed Houdini. "Got a weak section, he'll find it," his owner explained.)

Meanwhile, April, Karen, and volunteer Mary Ellen Moore were cleaning the large cow barn at the back of the farm. Only I didn't know this.

Murphy and I walked into the barn. "Where's April?" I asked the yellow mutt, who trotted toward the kitchen halfway down the aisle in search of his pal.

Five feet from the entrance, Rambo lay in the middle of the aisle, holding court per usual. Beside him stood Agent Forty-Four the turkey, gently pulling bits of hay from Rambo's wool. Potbellies Zoey, Charlie, and Ozzi were there, too; surprisingly, they weren't searching for food. They were simply there, enjoying the company of their friends.

I plopped down with them. Murphy did, too, right by my side. To my delight, the animals allowed us to enter their peaceful circle—and just *be* with them. No one charged over to beg for food; neither did anyone walk away because a dog and a human had entered his space. Hannah, Rambo's woolly pal, the sheep found in a Queens cemetery, strolled over to nuzzle Murphy the way she always does. Mufasa the goat was with her. Above all of us, Maxx, the sweet old gelding who recently moved into Dino's stall, hung his big head over the four-foot stall wall, and there we were together: two sheep, a turkey, a goat, three pigs, a horse, a dog, and a human.

For a few precious minutes we sat . . . that's all. Miraculous simplicity.

But then Claude—the giant pink pig with the bad leg that earned him free-range status lest he be picked on by the stronger, more dominant pigs in the pig pasture—strolled in from the far end of the barn.

"Hey, big man . . . hey, sweet pig," I called softly to him. A few heads turned in his direction.

"Mmmph," he responded. And then he walked not into his stall the way he always does, but past it, straight toward us. He walked right up to Maxx the horse, his scratchy pink back nearly level with Maxx's muzzle, and he lifted his snout to Maxx and there they were, wet pink pig nose pressing into soft black horse nose. They stood there, pig to horse, Claude looking up intently, somehow knowing that Maxx would not lunge at him the way the horses typically do at the big pigs. The way Maxx, who truly can't stand pigs, always does.

Laugh, shake your head, call me anthropomorphic if you're skeptical or obtuse or disconnected or afraid, but I experienced what I experienced, and what I experienced on a cold winter day was pigs and turkeys and goats and humans and horses and sheep and dogs enjoying each other's company. Happy just to be.

Carpe the Diem

Dear Maxx:

There is nothing more that we can do. Five years you've had with us, and five plus thirty-two, the age at which you came, is thirty-seven. You're an old boy, Maxx. Very few horses get this many years.

Still, that fact is little comfort as I face your final day on Earth. How I love you, Maxx! I struggle hard to breathe deeply, to gather the strength that I will need to send you off with a smile on my face a few short hours from now. I hope I'll be able to do that; I hope we all will. It's what you deserve.

Like the rest in this menagerie who call CAS home, you came to us because you had run out of options. Our waiting list is long and getting longer, as are the lists at sanctuaries and shelters around the country. Opening a thousand more shelters wouldn't put a dent in the volume of animals desperate for a safe haven, desperate for someone to say, "You matter to me." All sanctuaries have to pick and choose, and when your guardian angel asked, "Who on earth wants an old quarter horse with bad feet and a bum leg?" how thrilled we were to be able to say, "We do." We were her last phone call, she said. She had run out of options.

Yes, old boy: you mattered to us! A posse of people tried to help you after cancer took your human. They didn't want her final day on Earth to be yours, too, and their efforts to find a haven for you were truly Herculean. How stunned they were when I said, "Yes. We'd be delighted to give Maxx a home."

And so you came to this place where the unwanted come—a place where animals roam acres of pasture filled with rich grass, where they have spacious barns in which to bed down for the night, where they see not contempt but kindness in people's eyes. Here, animals learn what it feels like to matter to someone—to *many* someones. In this sense, you were far ahead of the game.

To our delight, you were everything they said you were: smart and comical, trusting and unflappable. And while they didn't mention it, we humans all admired how accepting you were of the discomforts and intrusions of aging: the creaky bones, the missing teeth and soupy meals, the arthritic shoulder, the painful and recurring laminitis. When you coliced last summer, you were tubed *five times*—five times in four days, the vet ran a half-inch rubber tube through your nose, down your throat, into your belly, pumping you full of mineral oil to dislodge the impaction that just wouldn't pass through. Five tubings, four nosebleeds, but man, Maxx, you were a trouper. How very clear it was that you knew you needed help, clearer still that you weren't ready to go. When you finally pooped (six times in a single afternoon), we were overjoyed! Ebullient! Yes, it was selfish: none of us were ready to say goodbye.

You were so much more than "horse," Maxx. You were teacher, friend, playmate. Yes, playmate! With Hannah and Ozzi and Tucker the cow and madman Clarence the miniature horse and Nutmeg the Invisible Hen in our midst, we certainly have our share of characters! But none like you. Whether you were stretching your neck out to grab the edge of the wheelbarrow with your teeth to drag it closer for inspection, lunging at poor unsuspecting pigs passing by, or play-fighting with Ted through the stall window, you entertained yourself constantly, and in doing so, entertained us. There's 500-year-old Maxx galloping at full-tilt the entire length of his pasture to ensure that Noah and Ted, the other two geldings, have no chance with *his girls*. There's Maxx with his head in the garbage can and oops . . . there's Maxx in the kitchen. There's Maxx, his head centered on a human's back, flinging it up and sending her flying across the aisle. Again. Again. Again. There's Maxx removing his support boots. And the final image: there's the brown streak of you, Maxx, dashing past Julie's office. The body ages; the spirit needn't. That's what you modeled for us.

How quickly you became our "starter horse." Yes—that was your job, Maxx! The tentative volunteer could enter your stall without fear, and when she was ready, you could reassure her as, instructed by humans who'd done this thousands of times before, she pulled the noseband gently over your muzzle, slipped the nylon strap first over your right ear, then over your left, then clicked the buckle and voila! Your halter was on, and she could lead you to your pasture, and you would be so very gentle, knowing, perhaps, that she was afraid.

You were the only one atop whose fanny I would cross my arms, resting my cheek on the crest of your tail, saying, "I'm going to kiss your ass, Maxx." I did, over and over, and you'd stand motionless, and together we'd prove the point that character is far more important than species in determining "how to be" around an animal—what behavior it safe, what's off limits.

These lessons and dozens more.

～ᗐ

"Heather's here," says April over the walkie-talkie, and from various stations around the farm, staff and volunteers gather in front of your stall to say a gut-wrenching goodbye and honor you whom we all love. April and Walt, Allen and Alex, Abbie and Troy, Kathy, Chris, Anthony, Donna, Melissa, and Amy.

Heather injects your fetlocks with a nerve blocker so that you can leave your stall. It's simply too difficult to get a body as big as yours through a three-foot door. If you can walk out comfortably, it will be so much easier after you're gone. And you do walk— exactly seven steps and no more. I walk backward in front of you, and when you stop, I drop the lead rope. It is your right, not mine, to decide which step is your last.

You stop, not surprisingly, in front of your friend Ted's stall.

"You done, old man?" I ask.

Our beloved vet Heather O'Leary gently explains what's about to happen for those who've never been present for euthanasia. Other than her gentle voice, there is only muffled crying. We are steeling ourselves. We do not want this. *I do not want this, Maxx.*

"Are you ready?" I ask, looking into your knowing eyes.

I suspect I will tell the story of what happens next for as long as I live. You turn your head to the left, stretching your neck as far as it will go, to look back at your human friends who stand on that side. Your gaze lingers a moment on each of us, then you bend your neck around to the right, looking behind you now at Walt, at Chris, at Abbie. Such profound clarity in that unexpected final act of saying goodbye.

Thank you, Maxx.

You crumble to the ground as the anesthesia takes effect. Gently I lift your big head, scooting under it until it is in my lap. I rub your cheek as you fall soundly asleep. Heather asks if we're ready; she injects the drug that will stop your heart, and you are gone.

Sometimes, being silly, I'll say to Murphy, "Come on, mutt, we've got to carpe the diem!" and we'll dart out the door to whatever adventure awaits us. You did that, too, Maxx. Up until the very, very end, painful shoulder, sensitive belly, and bum feet notwithstanding, you carped every single damned diem—no, every single damned moment—you were given. It was your gift, your abiding lesson for those of us who remain.

Seize the day. I'll do it in your honor.

The Slaughter Truck

I saw one today.

"Don't take Route 81," my Dad cautioned when we discussed my route home from Nashville, Tennessee, the final stop in a six-city book tour and a great excuse to see my ninety-three-year-old grandmother and some other beloved relatives. "It's a truck route. Scary as shit to be boxed in by three tractor trailers going seventy-five miles per hour."

I considered Dad's advice. I didn't relish the idea of driving nearly six hundred miles, much of it mountainous, surrounded by trucks. But the other option, driving *way* east via I-64 then heading up I-95, would add close to a hundred miles to the trip. So at 6 AM, I said a teary farewell to my aunt Beverly Ann, her husband, Frank, and mutts Bailey, Sammy, and Levi, then pointed the car toward Route 81.

Around 1 PM, with the sun high in the sky, the day warmed. I opened the windows . . . and that's when I smelled it—a slaughter truck, climbing the hill in the slow lane as I approached it on the left.

I don't often travel long distances via interstate highways, so it's rare that I encounter these deathmobiles. I've seen chicken transport trucks jammed so tightly with crates of chickens that many have suffocated by the time the animals arrive at the slaughterhouse for their barbarous deaths. Long before I began the work of trying to raise awareness of these delightful beings' sentience and the depth of their suffering, I wept when I passed the trucks. Aluminum boxes on a flatbed, rows of oval holes cut into their sides for ventilation. But that's all—that's the single accommodation for the animals, and that's done only so they won't die en route to the place that will slit their throats, dip them in boiling water, rip out their feathers, neatly slice off heads and feet, and clean and package what remains of their battered bodies for humans who would be far healthier if they did not consume their poisoned flesh laced with pesticides, antibiotics, and growth hormones.

Today, I did more than weep. You see, I know who these animals are now. I know that they're so much more like I am than I ever would have believed. I know that each one is individual, each one is unique, and I know that each chicken, each pig, each cow, duck, and turkey that is grown and killed to feed us has an emotional range that is probably quite similar to mine.

In my first book, I recalled with delight the life and lessons of one chicken named Paulie. Paulie was the barn peacemaker, a frequent passenger in my car (I usually insisted that he ride shotgun, though my lap was always his preferred seat), a good friend to my dog Murphy, and our regular companion at lunch. There have been other chickens, too, birds so full of quirky personality and a desire to communicate that one swears they really have vocabulary *if only we could understand it.*

Today's truck was stuffed with pigs.

Stuffed so tightly that what I looked at through the ovals was just a solid mass of pink. No doubt snouts were jammed into rectums and sharp hooves into tender skin because the goal, of course, is to smash as many bodies into the compartment as can fit—whipping them, shocking them, beating them—whatever it takes to get them on that truck. And friends, pigs are smart and pigs are sensitive and pigs are strong . . . they don't go willingly.

So as I passed this truck carrying animals I know to be uncannily "human," one pig caught my eye. He looked at me through the oval hole, and the look shared more than words ever could. In that moment, he was every animal ever grown to feed humans, and in this helpless, hopeless moment, he was asking a simple question: *Why?*

A wail emerged from my body. Not just tears. An uncontrollable wail—I could not stop it—coming from a deeper part of me than tears ever have, and an apology to that pig, and to all animals on behalf of my species.

I will return to CAS, where I will hug my pink pals Franklin and Policeman and Babe, and they will love me right back, with smiles and happy grunts and snouts rubbed into arms and cheeks, so that within a moment, I'll be happily as muddy as they. And I will wonder about my good fortune to be born a human and not any other kind of animal.

Ozzi's Bed

You'd think that a heated stall, a fluffy pile of fresh hay, and his best friend Mabel to snuggle with would provide enough bedtime comfort for Ozzi the talking pig to settle in happily at the end of each day. Mind you, Ozzi is a free-range potbelly pig who takes full advantage of his privileged status. In his relentless search for food, little Ozzi probably logs a good mile each day on his six-inch legs in search of tidbits dropped by horses or ducks or chickens. So surely after a hearty dinner, Ozzi should be ready to settle in like the rest of the pigs.

But he isn't.

Yesterday, we watched Ozzi sneak into the feed room (well, as much as a pig can sneak), grab an empty fifty-pound feed bag, and haul it back to his stall. Stepping on the thick paper bag, he grabbed a corner, tore, spit it out. Grabbed another corner, tore, spit it out. When his effort occasionally failed, Ozzi would simply hold the bag tightly and madly shake his head until another section gave way. Eventually, he'd gleefully shredded the entire bag, then carried it piece by piece to his preferred spot against the wall, arranging his bed with care. "You'll sleep well tonight, Oz," I said to him. Well, yes he would. The bag was just the beginning.

Ozzi stepped back into the aisle and fewer than ten feet from him, Murphy lay in the aisle, chewing on a blanket he'd retrieved from a donated armoire where we store all things soft and fuzzy. At CAS, we refer to anything made of cotton or fleece as "Murphy's towel," because sooner or later, whether it's folded neatly in the linen closet or hung on a stall door or worn by a human to keep out the elements, Murphy will spot it, take it, and have his way with it. My towels all have holes in them; my barn jackets have holes in them. Even though every child who visits CAS practices saying "I'm not your towel!" as soon as he's introduced to the yellow mutt, hundreds have been pulled down the driveway by their fleece coats, laughing hysterically.

A pig is even more driven. Whether it's being the first to be served breakfast or coming outside to root and play and get into trouble, when a pig makes up his mind that he wants something, watch out. There's no willful like pig willful. Pigs want what they want, instantly and with every fiber of their being.

A lovely quilt is balled up between Murphy's front legs, and he's gnawing blissfully on it, oblivious to the goings-on around him . . . oblivious even to the 200-pound black tank standing a foot in front of him, eyeing the quilt. An unclaimed section of the quilt extends out beyond Murphy's front paws. Ozzi stands there, thinking, and then he grabs the quilt.

Now, my dog has always excelled at tug of war; I sit on the hardwood floor with one end of a towel, and he happily drags me around the house, one good yank at a time. The *point* of the game is for Murphy to have fun "winning."

This is a very different scenario. Ozzi's got a good mouthful of quilt. He backs up. "Cool! Tug of war with a pig!" Murphy responds. He stands, and with all the strength a still-youthful thirteen-year-old dog can muster, he pulls back. Yet it takes but a single tug . . . Murphy is no match for Ozzi. Though he's uttering his fiercest growl and pulling with all his might, Murphy might as well be a chipmunk right now, because Ozzi is sailing backward and into his stall with his latest prize. Tomorrow, he'll place an order for a foam mattress.

"Good job, Murphy!" I exclaim. "Good tug of war!" I kiss my yellow dog's nose and, still chuckling at Ozzi's antics, go to the linen closet to retrieve another towel for Murphy.

Devotion

My criteria for a car are simple and few: it must be all-wheel drive, it must be fuel-efficient, and most importantly, it must be a comfortable ride for my beloved pal Murphy, Catskill Animal Sanctuary's director of canine pursuits, who still wants to accompany mama *everywhere*. Of the three, the third criterion is the deal-sealer. So imagine my delight when I pulled into Prestige Toyota, cash in my pocket, and discovered the certified used car of my dreams. A very few minutes later, the deal was done, and Murphy and I pulled out in our 2005 Toyota Matrix—the tiniest hatchback "wagon" I've ever seen.

What I especially like about my car is that both the back seat and the passenger seat fold down absolutely *flat*. Murphy can lie on his monster mattress and literally rest his chin on the back of the passenger seat and be *right there* with me.

These things matter when one is devoted to an animal.

I'm not more than a hundred yards out of Toyota's parking lot, thinking about the concept of devotion, when my cell phone rings. It's my assistant, Julie.

"Can we take two five-day-old blind lambs?" she asks. "I can't reach Abbie."

Abbie is our animal care director. It's her day off. Generally, I like for both her and me to say "yes" to incoming animal emergencies. Not to consult with her, the person who will be doing the primary caretaking, is unfair. However, this time, I hear myself saying "Oh . . . of course we'll take them!" as my eyes well up with tears.

Murphy and I make two stops. When we pull up to the barn, Karen and Abbie (on her day off) are huddled in a heated stall around two five-pound sheep, jet black, no markings.

"Oh, my," I say to them.

"I know," says Abbie. "Can you stand how cute they are?"

"Who needs kids?" we tease Troy, devoted dad of two young children, who has just poked his head in.

I don't know why, but we give the two lambs Irish names: Patrick and Phineas.

We will follow the same feeding schedule that we've followed with other orphaned animals: feedings every two to three hours, including throughout the night. Only this time, I'll do just the 5 AM feeding. That's the time when Murphy and I begin our day. It will be Abbie and Karen's job to do the 8 PM, 11 PM, and 2 AM feedings, just as it's their job to stay up late with a sick horse, stay late when the vet, who's supposed to arrive at 3 PM, finally pulls up to the barn at 5:30, or what have you. It's their job to consult with multiple vets about a troubling diagnosis, to know when to ask for my input, to make judgments about when it's time to send a sick animal to the hospital—all this and more *on top of* the "routine" parts of their respective jobs. For Abbie: creating optimum diets for 250 animals; conducting thorough weekly health checks; doing "critical care" each morning on sick, injured, or chronically ill animals; trimming beaks, hooves, and nails; scheduling an array of health-care providers; maintaining thorough medical records; keeping a watchful eye on herd dynamics and adjusting living arrangements accordingly. You get the picture. Karen's job, in addition to all of the above on Abbie's days off: setting up feed twice a day, organizing and supervising staff and volunteers to work as a team in the interest of the animals, and working with withdrawn horses to build their trust and confidence. Asking these two to work around the clock is a tall order.

The two young women tease me about setting them up for overnight shifts. "*Thanks a lot!*" Abbie says, but as she says it, she gives me a hug. Karen says simply, "It's okay, Kathy. It's part of the job."

It hits me like a ton of bricks: this is what devotion is. And my next thought is this: *that's exactly what makes this staff the staff of my dreams.* Devotion. Every single one of them is devoted to this special place—April, Julie, and Alex, here from the early days; Troy, Abbie, Karen, Michelle, and Caleb, all here within the last two years but such extraordinary additions to our family. It's devotion that manifests itself in myriad ways large and small, but what it looks like from where I sit is a group of people giving the very best of themselves, to the animals, to me, and to the organization, every single day.

Granted, it's taken us a while and plenty of growing pains to get here. I've had to learn patience—hardly my long suit. I've had to learn that I can't just set high expectations: I have to support the staff in achieving them. Every single one of us, in fact, has had to grow in ways that have sometimes been uncomfortable, but we have grown, and Catskill Animal Sanctuary is the beneficiary.

I'm not sure what I've done to deserve a staff such as this, but I am sure it's time for me to publicly acknowledge how grateful I am. They are precious gifts. Thanks to them, Catskill Animal Sanctuary is a place of uncommon beauty and grace, a mirror for the world as it should be.

I close the stall door as Abbie and Karen are offering the lambs their first bottle. When I hop back into the car, Murphy's head rests on the back of the passenger seat. Devotion: it's the theme of the day.

The Little Horse That Could

"He's too weak to walk," Stephanie Fitzpatrick of the Dutchess County SPCA told me before she pulled into the driveway with our latest rescue. She wasn't exaggerating—little Andy collapsed when he got off the trailer, and then fell again in the driveway. Looking at him now, this little white horse with hundreds of brown spots, I can't believe he is alive. He's nothing but bone. But he *is* alive . . . and that means we've got work to do.

Andy's owner runs a "nurse mare" operation that rents out mares for their milk. Unless they're to be used for breeding, male horses serve no purpose. Like male chicks in the egg industry, male goats and cows in the dairy industry (prior to the invention of veal), or male horses in the industry that produces estrogen replacement drugs from the urine of pregnant mares, they are considered by-products and are discarded by any means the business owners wish. Baby chickens, for instance, are suffocated, crushed, or sent to gas chambers by the thousands by the egg industry; the industry has no use for the birds who don't produce their product.

Andy, too, was "useless," and he was dying from starvation.

Coming into work the morning after Andy's arrival, Alex burst into tears when he saw our new friend. It's everyone's reaction. As a three-year-old Appaloosa stallion around fourteen hands, Andy should weigh eight hundred pounds or so, but at most weighs half that. If you're of normal weight, picture yourself with half your weight missing: you'd be skin draped over skeleton. That's Andy. The vet labeled him "below 1" on the Henneke Scale, a system used by veterinarians to represent horses' body conditions, with 1 being the lowest possible number on a 9-point scale. A Henneke 1 depicts a horse in danger of dying. In many previous cruelty cases involving over a hundred horses total, we've had only seven horses considered by examining vets to be Henneke 1; none were nearly as debilitated as Andy. Surely nothing but will has kept him breathing.

I stand in front of his stall looking in at the little horse, who looks part Dalmatian, part deer, part horse. His legs are pencils; his rump higher than his shoulders due to absurdly long rear legs. He strains to poop, but passes water instead of manure. Later, his urine is the color of dark chocolate. We notice a thick pus-like substance running down his back legs. His rectum has prolapsed; it hangs outside his body, a raw and painful mass. We wrap his tail in an ace bandage to prevent further irritation and infection.

"Do you think he'll make it?" I'm asked a bazillion times a day.

"He made it here alive," I say to everyone who asks. "This is easier than what he's been through."

Of course, I know that Andy's recovery will be anything but easy. In fact, I am concerned that he is in organ failure. But Andy is one of those animals with whom everyone has fallen instantly, hopelessly in love. We love him for his courage, for the light and brightness in his eyes in spite of what he's been through: chronic starvation for probably his entire life. He is going to fight; we will fight with him. Part of that fight is surrounding him with hope and positive energy. So volunteers and visitors hear that Andy is doing beautifully, even when he's struggling.

A few days after Andy's arrival, I enter his stall and sit with him for a long while. He is lying down. His sides rise and fall hard—even the act of breathing is difficult. "*You did it, little man!* You're alive, little trouper. Good job!" I whisper to him as he rests. A lesser spirit would have given up long before now. He sits up and looks at me without malice. No anger, just exhaustion and quiet determination. He nibbles my knee. "Hey, silly boy," I say to him. When he decides to stand, I back away to give him room. Left front leg stretches out, then right. Andy struggles mightily; his entire body trembles.

"You can do it, Andy!" I encourage him. "You're such a strong boy!"

Andy is shaking violently, but he has made it. He is standing and I am weeping; it's the first time I've seen the heroic effort required simply to stand. He leans against the stall wall for support. I lean into him with my full weight, my back just behind his shoulder, thinking that perhaps the extra support will be enough to help him last a few minutes longer. Even if it doesn't, I want everything we do to convey that we're with him in this. His sides heave from the exertion; his tiny legs tremble. After no more than two minutes, it's time for me to step aside.

"Okay, *Andrew*! What a great job!" I praise him. Andy folds his front legs under him and collapses with a thud.

We've triple bedded Andy's stall with three heaping wheel-barrows of shavings and frequently refer to the "Princess and the Pea." But even with this generous bed, lined with rubber mats, Andy has large raw abrasions on his hips from where the bones rub the floor. We treat them and add more shavings.

I lie down in front of Andy's head; the two of us form a T. I rub his cheek, his neck, and his forehead, and for a few moments, he nibbles my shirt. Within minutes, he's sound asleep. He sleeps for most of the day, of course. In addition to a medically supervised diet and a growing fan club, quiet, comfortable sleep is what he needs in order to heal.

Andy's owner has been arrested and charged with cruelty. While her fate plays out in the courts, we will do everything we can to heal this lovely boy, whose ebullient spirit has captivated us all. After just four days, he already wobbles on unsteady legs to his front wall, leans out, nibbles a cheek, pulls a hat off an unsuspecting head. Too weak to muster a full whinny, he calls out a pitiful "hello"—all breath, no sound—to everyone he sees.

He's a trouper. And we're rooting for him.

❧

October 17: a month since Andy's arrival. He's still a mess, but at least he's not contagious. That's a piece of very good news. We put away the bleach baths into which we've been dipping our feet. It feels good to touch him with my skin rather than with a gloved hand.

Andy is still a bag of bones, still extremely wobbly on his feet. He's gained perhaps forty pounds, but is still so skeletal that there's no point trying to use a weight tape, a measuring tape used to esti-mate horses' weights that is supposedly accurate within twenty-five pounds. Veterinarian Heather O'Leary says his condition will skew the results.

One has to use extreme caution putting weight on a debilitated horse: small frequent meals of hay only, gradually increasing the amount, decreasing the frequency, and eventually adding small quantities of grain. Andy's case is especially sensitive—his starvation was chronic; his organs so stressed and compromised. Like so many who've come before him, Andy is on a special refeeding program. Yet despite the measured discipline of Andy's weight gain regimen, he's getting far more food than he's ever received. Nearly as soon as he finishes a flake of hay, another one is on its way—an alfalfa mix; lots of energy, lots of calories. He eats with relish. Soon, we'll add grain to his diet.

Just as important for Andy's healing, however, is the attention he's receiving: the kisses, the gentle words, the grooming, the encouragement. It's rare that I walk in the barn and see Andy alone. Normally, one or more of his sizable fan club are just outside his stall, either stroking his face or allowing him to chew their coats, their hats, their hair. He relishes the attention!

If Andy could speak our language, he'd probably tell us delightedly that he's the first CAS resident to receive a daily sweeping. Mind you, I don't make a practice of sweeping horses' entire bodies. But when one day I simply could not sweep the aisle outside his stall because he insisted on grabbing either my shirt or the broom handle, I turned his playfulness into a game.

"Okay, little man, you want the broom? You can have the broom!" I said, and brushed his entire face from ear to muzzle. Andy grabbed on, and we played gentle tug of war until a perfectly good broom was about to be destroyed, at which point I dropped the handle, went to him and opened his mouth, forcing him to let go. I entered the stall, and I swept Andy's neck, his back, his rump. Andy stretched his neck out as far as he could, indicating how much he liked this game!

This work—if you can call it work—is one of the greatest joys of what we do. Any of us can tell you what a privilege it is to participate in the healing of a broken animal—to say with every word, touch, gesture, "You're safe, sweet one. You made it." Andy is far and away the most exuberant spirit I've ever known and thus makes our job that much more rewarding. But whether joyful and outgoing or fearful and reserved, all hearts deserve to sing. What an extraordinary gift it is to be one among those who offer that opportunity.

Andy is definitely standing more solidly on his pitiful pogo stick legs. His manure is generally solid; his rectum is staying inside his body. And it could be wishful thinking, but I don't think so: I swear his whinny is getting stronger. It won't be long before Andy's ready for his first real adventure: a walk outside to explore his new world at Catskill Animal Sanctuary—a far cry from the nightmare that cruelty investigators found him in.

～〜〇

One hundred and twelve days. That's how long it's been since Andy arrived. That's how long it has taken Team Andy, comprised of veterinarian Heather O'Leary, master farrier Corey Hedderman, former Animal Care Director Walt Batycki, and dozens of others offering love, encouragement, and time, to bring him to this day. One hundred and twelve days. But we've arrived, and he's ready. Andy is going outside.

"Come on, little guy," Walt says to him as he attempts to put the magenta nylon halter over Andy's head. It's a real challenge, of course: Andy wants to chew the halter. Andy wants to chew *every-thing*. He grabs the halter quickly, again and again. Finally, I hold his head while Walt slides the halter on. Though we've been working on this, it's clear we have a long way to go. (Andy is now four years old—a young adult—but everything about him is still a baby. I don't know whether it's simply who he is, or whether, deprived of a mother and a herd of horses with whom to socialize, Andy was never taught how to behave, never taught which behaviors were acceptable and which ones were off-limits. Certainly no human ever taught him. Regardless, his mischievous nature endears him to everyone he meets.)

"Andy, we're going to take a walk!" I exclaim. Walt clips on the matching magenta lead rope.

"You ready, fella?" he asks. A gaggle of staff and volunteers has come in to witness Andy's big moment.

"I can't believe he made it," Sarah, a volunteer, offers. "I can't believe he made it to this day."

Tears fill my eyes and I pause for a moment. "This is a good one, guys," I say to the group. "Congratulations."

"Aw, shucks," Walt says. "It's been a blast working with an animal who tries to eat my face every single time I go in the stall."

It has been a blast . . . for all of us. Indeed, it has been an honor.

"You ready, fella?" Walt asks again.

"Here you go, boy," I say as I open his stall door wide. We stand, our arms around each other, and watch in awe as the Little Horse That Could steps out of his stall.

All Andy does on this first day is exit the barn, circle it, and come back into his stall, walking perhaps a total of three hundred feet. But every single one of us witnessing this moment recognizes it for what it is: a miracle.

~꙳

I enter the barn from the side near my house, and as I glance down the aisle, Andy is doing his "Oooh, it's recess! I get to go play!" dance as volunteer Kathy Keefe leads him from his stall.

"Hold on, Mister!" she says, and as I approach them she adds, "You know, somebody needs to tell this boy he was nearly dead nine months ago. Apparently he's forgotten."

I turn to walk outside with them. It's not that I think Kathy will have trouble. She can handle an agitated horse just fine. It's that I want to witness what's about to happen, for even at a place where joy abounds, Andy's first moments in the pasture are always worth the price of admission.

Andy shares a field directly behind the barn with five other horses: geldings Noah and Bowie, and mares Callie, Hazelnut, and Crystal. These five stay in the field round-the-clock. We call them the Kind Group. There's not an attitude in the bunch. Bowie and Noah—each of whom had his own stall in the barn until a) their physical and psychological healing was complete and b) their behavior suggested they'd be happier as outside horses—are the newest additions.

I open the gate and watch as Andy chews on the sleeve of Kathy's jacket and does a little jogging in place. He's like this each morning: antsy. She leads him in and releases him, saying, "Go get your buddy, Andy!" In just two strides Andy is in a full out gallop, charging across the hill to his best friend, Bowie, who lifts his head as Andy approaches. As Kathy returns to the barn and her long list of chores, I follow the spotted imp.

Andy's pal Bowie, a lovely quarter horse, arrived just four months ago. He was two hundred pounds underweight and cowered when anyone entered his stall. While he was far mellower than Andy, what the two horses shared was a strong desire to get past their history. That plus youth: while the older mares found Andy's insistent personality annoying, and Saint Noah tolerated him, Bowie, just five years old himself, found in Andy the perfect playmate. Their friendship was nearly instantaneous.

Andy has found his pal, and for a long moment, the two young boys stand head to head, breathing each other in. Then, without warning, Andy does a funny little scoot backwards, then a quick scurry around to the right until he's at Bowie's rump. He nips his friend playfully on the fanny, and the games begin. For the next twenty minutes, they play as young foals play—cavorting and kicking, bucking and rearing, charging through the field. Bowie stops short and Andy stumbles. No matter. He's back up in a flash, grabbing Bowie's tail with his teeth.

One year after the Little Horse That Could arrived at Catskill Animal Sanctuary, he still looks part horse, part gazelle, the likely result of inbreeding and chronic, extreme malnourishment. His back legs are a good four inches too long; his hocks jar violently to the side each time he takes a step. He'll never win a beauty contest. But that doesn't matter. What *does* matter is that Andy recovered. Class clown. Nudge. Annoying little brother. Victor. Teacher. Hero. He's the little horse that could . . . and he did.

❦

"Come on, mutt," I say to the yellow dog. "Let's go see the animals." As much as they enjoy our nighttime visits, I'm the needy one tonight.

I was just finishing up a phone conversation with a relative whom I'll call "Sam" when Sam said, "Just one more question." I braced myself, silently cursing that I hadn't successfully maneuvered an end to our chat.

Sam and I have a long history that binds us despite many profound differences. He is a conservative through and through. And though he'd defend himself by saying "one of my best friends is black" (which is actually true), I consider Sam to be racist. I often confront him on this, though in general, we've learned to avoid topics where we can find no common ground.

"Whaddya think of our president?"

"Sam," I said, my voice steady. "Why are you asking me this?"

"Don't tell me you support him!"

Oh boy. Here we go.

Despite similarities in other areas, Sam and I are political polar opposites. It's not this that I mind so much. What I mind is that a perfectly bright man and an essentially kind one shows a shameful ugliness and an utter unwillingness to listen to intelligent argument during any kind of political talk.

"He's Muslim . . . and a *socialist.*"

"Actually, Sam, he's neither."

"Barack *Hussein* Obama—what in the hell kind of name is that? It ain't a Christian name! Shit, the thought of Hillary in the White House was bad enough—you know, she's a socialist, but this guy, I tell ya . . . I'm thinkin' of moving to Wales."

Wales?

"I've been doing some research at the library, and if that crock o' shit health care bill passes . . . "

"Why *Wales?*"

"Wales looks like pretty country. There's no socialists and no towel heads."

"OK, Sam," I say, refusing the bait. "If health care passes, you leave. If it doesn't, I leave."

"Oh, yeah?" he says, delighted with the opportunity to continue this nonsense. "Where you gonna go?"

"Canada."

"*Canada*? Ain't no more Canadians in Canada. It's all A-rabs— towel heads everywhere you look."

The vitriolic diatribe continues until I am able to say, "Sam, I've got to go check the animals."

"Okay, girl. You tell them pigs I said hello," he says, then hangs up.

In the bedroom, I pull on one sweatshirt, then two. As I do so, I put words to the feeling that has been creeping in over the last few weeks: I am utterly exhausted.

This wonderful Earth, her needs ignored by a species that believes itself entitled to use her up without consideration of the consequences, is dying. Our economic system is weakened and our debt grows by the minute. Obama, the "yes we can" candidate who stood for a fundamental break from business as usual, is sending 30,000 troops to Afghanistan. Among people like Sam, anti-black rhetoric has been replaced by anti-Arab venom. And on and on it goes. Far more than the work of running Catskill Animal Sanctuary, the state of the world is knocking the wind out of me.

❧

"Let's go see the animals!" is always Murphy's cue. He makes a quick detour to the bathroom, steals a towel hanging beside the tub, then leads the way through the darkness to the big barn.

I pull open the huge sliding doors and step inside. The long barn aisle—ten stalls on each side—is gently lit by several two-watt nightlights. For a moment, I simply stand. At the far end, Bobo the blind horse whinnies—she knows that treats are coming and as always, she wants to remind me that she should be the first to receive them. On the left, Lexie, an old mare spent from her years of producing babies for her owner, munches on hay. Her broodmare belly sags; her body is worn out. On the right, Zen the goat jumps up, resting his front hooves on the four-foot stall wall, stretching his neck as far as he can in hopes perhaps of a kiss, but mostly for what he knows is in my pocket: alfalfa treats.

"Kiss first," I say to Zen, and he obliges with a cool nose touched to mine. "Here, little man," I say, presenting the treat.

Three stalls down, Beacon, the ancient potbelly who has stolen the heart of everyone who meets him, snores beneath mounded blankets. If I were to open his door, all I'd see was a two-foot pile of blankets that the little pig has painstakingly arranged before very carefully crawling under them for the night.

"Hey Rambo," I say to the great sheep, our guardian, resting in the aisle on his fresh bed of hay. I bend down to rub the space between his horns; he blinks slowly in gratitude.

Stall by stall we proceed, a kiss and a carrot or alfalfa cookie for each resident. As I linger with each one, Murphy lays in the aisle, gnawing on his towel.

I peer into Andy's stall. As usual, he is stretched out flat in an absurdly deep bed of shavings. His recovery has been miraculous; a combination of good care, expert farrier work, and sheer will. Lesser spirits would have given up. Not little Andy. Still, after a full day of running amok in the pasture with his best friend Bowie, Andy is spent, and just like he does at the end of every single day, he has folded himself onto the floor and stretched out for some well-earned sleep.

"Pooch," I request of my dog, "May I have your towel?"

Murphy is completely uninterested in releasing his prize, so I explain that I'll be right back, then head to the kitchen to retrieve another from the armoire.

"Hello, little man," I whisper to Andy as I open his door. "You've had a day, haven't you?" Andy looks up but does not budge from his prone position. Murphy follows me in, dragging his towel.

I lie on my side so that my entire body stretches out against Andy's neck and back, and drape my left arm over his neck, propping my head up on my right hand. Such a connection would frighten many horses, as would a dog lying beside them. In fact, as prey animals, many horses never sleep prone the way Andy does. My sense is that it's because doing so exposes their belly and vital organs. But Andy? Andy stretches out flat many, many times a day. He lets go; he gathers strength. He recovers from a day that has challenged a body that's still in recovery from a lifetime of neglect.

Right now, with my body pressed next to him, his breath is deep and slow. For a few minutes, the three of us lie like this: Murphy lies at Andy's flank, gnawing with gusto on his towel. The miracle horse rests as I am stretched out against him.

Slowly, I sit up, scooting around to Andy's head. "Good boy, Murph," I praise the mutt, and then, looking down at Andy, I ask, "Andy, would you like a pillow?"

I take the huge white bath towel and fold it in half, then in quarters, finally in eighths. I place the towel in the middle of my lap. Scooting forward until my left knee touches the top of Andy's head, I slide both hands under his head, positioning them so I can manage the weight. Andy lets go, releasing the weight of his head into my hands that lift him enough so that I can scoot under him, then gently place his head on the fluffy towel.

In under a minute, the little horse's eyes are firmly shut. He is soundly asleep, his head in my lap. How grateful I am, especially tonight, for my little towel head.

～◯

And then one day, Deborah Smith-Fino showed up.

"I'd like to adopt two horses," she said. "One for me, and one for my husband."

Deborah, returning to the farm several times, proceeded to fall in love with Bowie; her husband with the alpha mare Athena. Now we *certainly* understood their choices! Young, striking Bowie—with his bald face, one pale blue eye and the other half brown, half blue— was another survivor of horrific abuse, and when Deborah appeared, Bowie was just learning to trust us. Deborah was smitten. Strong, confident, outgoing Athena was a no-brainer for Deborah's husband Eric. The couple's adoption application was approved pending completion of their barn, and they got to work to finish it.

Fast-forward to a conversation between Deborah and me in which I mentioned that Bowie and Andy were best friends who grazed nose to nose, charged around the field together like two young boys should, rested chins on each other's rumps. Given their individual histories, I was loath to split them up. Noah, a far more mature horse, would bond easily with anyone; he actually after a time found Andy rather obnoxious. But youngsters Andy and Bowie were attached at the hip. Given their histories, I was very reluctant to split them up.

A few days after that conversation, Deborah appeared in the barn. She watched as a tour stopped in front of Andy's stall and Andy tried to chew everyone who came within reach. She watched as I swept Andy with a broom. He still loved the game. She heard stories about how Andy falls sound asleep with his head in my lap, and how he allows me to stretch out fully on top of his prone body.

When I continued down the aisle with the tour, Deborah stayed behind with Andy.

Two hours later, I dropped off my final group back at the Welcome Hut, then returned to the barn to load up the truck for afternoon feed. To my surprise, Deborah was still with Andy.

"I want him," she said softly as I walked back into the barn.

"I don't want two animals who have been through so much to lose each other." And she said it again *after* she read through Andy's medical records—all thirty-some pages of them, including the prolapsed rectum, the liquid manure, his struggles to stand, the farrier work when he was far too weak to stand on three legs, the vet's description of his deformed body and how it might impact his health.

A few days later, our veteran (and vigilant) home inspectors Walter McGrath and Charlotte Mollo gave an enthusiastic "thumbs up" after their follow-up inspection of Deborah and Eric's newly completed barn.

The next Sunday afternoon, our hauler Corinne wound down our driveway with her fancy white trailer. Athena went on first, followed by Bowie. Andy was next. As Allen walked him up the aisle toward us, my tears began to fall.

"Do you mind if I load him?" I asked Allen.

"No, take him," Allen said with generosity.

I took him, our Little Horse That Could, and led him up to the trailer. Andy hesitated. Corinne was inside; I handed the lead rope to her and I stood right next to Andy, my hand on his neck. I'll never forget what happened next. Andy looked directly at me, and inside my head, I heard the words, "Are you sure I'm going to a good place?" as surely as if Andy had actually spoken them.

He was looking hard at me. He was shaking. He wanted an answer.

"Yes, my boy. You're going to a good, good place. I promise." I said to him. And he believed me. Andy took a moment to work up his courage, then stepped onto the trailer and stood right next to his best friend Bowie. We closed and locked the rear door, and they were off.

Thank you, Deborah.

And thanks especially to you, Andy. You're in our hearts, little man.

Just Another Day at CAS

Eight of us are climbing a narrow path up a Pennsylvania mountainside. Granted, this is Pennsylvania . . . we're not talking Kilimanjaro. Still, the climb is rocky and uneven, and our hands are full—we will likely be tired before the work even begins. Troy has a large dog crate balanced on his shoulder; Walt carries backpacks filled with rope, sheets, lead rope, and a first-aid kit. Volunteer Sharon Ackerman, an acupuncturist, carries water for the troops; the rest of us pass heavy crates between us. I carry one atop my head for a while, then pass it on to volunteer Vanessa Van Noy, who's taking a day off from yoga instruction. Lorraine doesn't even work at the Sanctuary on Wednesdays, but here she is, climbing with us on her day off, as is Friday volunteer Debbie Wierum. We are going to rescue nine feral goats. At least one is injured and unable to walk. Several are pregnant. All are emaciated.

"I see one of the black ones," says Gray Dawson, who is just ahead of me. His wife, Melissa, has left the group and gone ahead of us. It was Gray and Melissa who called us about the animals. Melissa has seen the goats five days in a row now at the top of her friend's vast mountainside property. Her friend can't fathom who they belonged to. "No one around here raises goats," he tells her. Per our instructions, the couple has been taking small amounts of grain to the animals, hoping to earn their trust.

The rest of us hang back in a pod, sitting very still, and sure enough, two pygmies come bounding down the mountain. Walt moves forward. In an instant, the two boys have lead ropes around their necks and are munching happily on grain. We lure them into the biggest crate, knowing we'll need all available humans to safely catch the seven remaining goats.

"We'll never get them," Melissa says dejectedly. It's the pygmies who have come to her several days in a row. Of those remaining, only one has come cautiously within her reach in preceding days. For the others, even though their emaciation is extreme, fear has consistently won out over hunger.

I am far more confident than Melissa, and though we don't talk about it, I suspect the rest of the CAS crew is, too. It's not like we've had a lot of experience rescuing abandoned goats from a mountainside. But we have taken in hundreds of animals who were originally (though usually not for long) terrified of us. We know how frightened goats move. We know they can spring and leap with remarkable speed. Our bodies know how to match their movements.

The injured goat and her very pregnant friend are huddled at the edge of a vertical rock face. There's a thirty foot drop to the craggy, boulder-filled forest floor. If a terrified goat leaps, she'll die; if a human slips, she'll die.

"Should I try to get behind them to drive them down the hill?" Melissa asks.

"Yes," I say, and as she does so, the rest of us very, *very* slowly encircle the two frightened animals.

A boulder juts upward between Lorraine and me—a boulder that a frightened goat will very likely use as a springboard to leap over our heads and out to freedom.

"Kathy," Lorraine says. I turn; she's found a long, large branch. We stretch to hold it between us, hoping an additional visual barrier will dissuade a fleeing goat.

We close in. We are all crouching, we are all quiet, we all utter soft words of encouragement. We are fifteen feet from the goats. Then we are twelve; then we are six. A few more feet and we can dive and catch them. But a goat is flying through the gap between Lorraine and me. I leap for her. We collide in a heap to the ground and are sliding downhill toward the cliff. Instantly, Walt and Troy are there, wrapping the frightened animal in ropes and a sheet to keep her both safe and contained. I sit up.

"Good tackle!" a voice from the left says.

"Thanks," I say, quickly dusting myself off.

Over the next hour, we continue this methodical work until just one goat remains. She's below us, backed into a corner below the rock face.

Troy scales down a steep slope, followed by Debbie. I can't see the goat.

"If all of us come down, can we trap her?" I ask.

"I think so," Troy responds.

One by one, we lower ourselves into the small cavern. This time, the goat can only move forward; the rock wall behind her is sheer and solid and completely vertical. We form a line and inch forward, our hands outstretched to close in the gaps. The goat, rail thin but pregnant, turns quickly in every direction, desperate to find a way through. She leaps but Troy leaps faster. They fall together and then Walt is there, safely harnessing her and wrapping her terrified, weakened body in a sheet.

It's clear we'll need to carry her out . . . and we do.

⁓

Three hours after we arrive, nine goats are safely loaded into a rented cargo van and are heading north to Catskill Animal Sanctuary. They will live in a large, hilly pasture filled with trees and boulders much like the forest they came from. Their babies will not fall prey to hungry fox or coyote. In this new home, they will have plenty to eat, warm shelter in the winter, and if they choose it, plenty of love from always-willing humans.

It's six o'clock when we return to the farm, but April, Alex, Kathy, Abbie, and Allen have all waited for us. We settle the goats into their deeply bedded stall; fresh hay and water are piled in opposite corners. Tomorrow, we'll separate out the pregnant moms, whose udders suggest that kids could arrive at any time.

We're tired but smiling. It's just another day at Catskill Animal Sanctuary.

Growing Up

It was December 2009 when the words came to me: Catskill Animal Sanctuary has grown up.

December 5. Weather.com is forecasting a high of fourteen degrees; other than my intrepid staff, who in their right minds are going to show up to help with routine daily chores or with today's particular priority: unloading 1,100 bales of hay from a flatbed truck? While we might have six volunteers *scheduled* today, I cross my fingers that we'll see even a couple and suspect I'll be adding a few more items to my essay "When Winter Kicks Your Ass."

But show up they do: Elena the pharmaceutical rep, kickboxer, and triathlete, who's been coming every weekend for six years; Brett, Simka, and Allison, the Bard College students; Kathy the purveyor of organic, fair-trade Catskill Mountain Coffee and devoted caretaker of Ted the draft horse; and Julie, the motor vehicles employee who lies awake at night worrying about the animals. It's only 9 AM. If our afternoon volunteers show up as well, we'll be golden.

"We must be doing something right for you guys to be down here today," I say loudly enough for all to hear. "I really can't thank you enough."

"Shoveling shit in zero degree weather is my idea of a good time," Kathy jokes from inside Lexie's stall.

From the far end of the barn, a bundled figure shouts, "Madame Director!" It's Frank, our neighbor from the top of the hill and the former owner of the land that is now Catskill Animal Sanctuary. "I see you guys could use some help this morning," he offers.

I have never seen Frank without a smile on his face. He's smiling now, and I realize it's going to be a good, good day. At seventy-four, Frank hoists two forty-pound bales of hay as if they were two toy blocks; once he climbs in that hay truck, regardless of whether he has six helpers or none, he doesn't leave until it's empty.

~

As fortunate as we are to have a huge posse of volunteers, it's the staff that bears the brunt of the grueling winters. In fact, I often refer to April, Julie, and Alex, each here from the earliest days, as "quiet warriors," and nothing tests their mettle more than cold, dark days on the farm. In the literal sense, Alex is about as quiet as a foghorn. But his work ethic is what I'm referring to. Alex doesn't stop. Quite remarkable, when you consider that his workday might include feeding a hundred animals, setting posts in sweltering sun, stretching fence, scooping poop, unloading hay, and feeding those same hundred animals again. He just keeps going, keeps smiling.

So does April. April packs enough energy in her ninety-five-pound body to power a hydroelectric plant. And though she's always been a worker, she has also become the leader we've needed her to be, modeling for volunteers both the work ethic and the attitude we require of them. "Bring nothing but love here," is what we say.

Julie, meanwhile, juggles everything from volunteer applications to newsletter design to all our membership and financial information to phones that never stop ringing. "You can *never* leave here," I joke with her, for we both know how exceedingly difficult it will be to fill her shoes.

These three veterans were joined in 2009 by Troy, Abbie, Karen, and Caleb. They, too, are peaceful warriors, all with a quiet determination to get their work done without any drama. It's a hell of a team, and there's no time when I appreciate them more than on days like today.

Winter is challenging for other reasons: it is when our hay costs skyrocket and animals most often arrive from cruelty cases, having typically been locked into a barn to fend for themselves. Thankfully, this costly time of year is also when our supporters shine, stepping forward each December to contribute to an ambitious end-of-year fundraising campaign.

Many donate $10 or $25, far fewer donate $500, $1,000 or more, and we have a dozen "even bigger than that" donors. Time after time, these combined gifts exceed the funding needed to care exceedingly well for some 250 charges. "Six thousand members can buy a lot of hay," my Dad says. And he's right: they do.

How fortunate we are to have gotten through this early phase of our growth: seven years of creating a farm, building an exceptional staff, recruiting a battalion of volunteers and devoted supporters from around the country. Getting here has required many heroes; I am grateful to all of them.

Last year, we called our end-of-year appeal the "CAS Bale Out." It was a topical reference that we hoped would draw people's attention, and it was a reference to *bales* of hay, our number one expense during the winter. The goal was to quickly raise $100,000, enough so that we would feel secure in saying to people desperate to place animals they could no longer afford to feed, "I'm so sorry you're going through this, but we have a space for your friend." Four weeks after the Bale Out was launched, CAS had raised $260,000. We were elated, and alone and in pairs, in groups and flocks, needy animals came to CAS.

This winter, a longtime donor to Catskill Animal Sanctuary came forward with an extraordinary offer to help us "fast-forward" our dreams. I had told him about our long-term plans to create a vegan cooking program, a summer camp for children, and a permanently expanded rescue program. While I knew he'd help, I didn't expect his response. "I'll match as much as you can raise, up to a million and a half dollars," he said. We truly had a once-in-a-lifetime opportunity to fund our future. We set two deadlines: December 31, 2009, for the initial push, and December 31, 2010, as the final date of the matching campaign.

Two weeks into the "Million and a Half Match," Murphy and I climbed the hill to the mailbox. I opened it and peered in at a six-inch stack of envelopes. Halfway down the hill, we plopped down on the ground. Murphy grabbed a stick the size of a small tree, and I opened the first envelope. It was a check for $10. The next dozen or so were for $10, $25, $50, or $100, then there was one for $5,000 with a note that read, "We're so excited for you and the animals! Happy holidays!" Most of the checks were local, but over thirty came from out of state: Washington, California, New Mexico, South Carolina, Florida, Tennessee, Virginia, Vermont.

Julie would tally our online donations, which typically equalled or exceeded those arriving by mail.

One delirious chew after another, Murphy crunched his stick down to nothing: he chewed off a chunk, spat it out; chewed and spat. It took half an hour to make my way through the stack to the envelope I saved for last. It was addressed in the uneven scrawl of a hand that's no longer steady. Two dollar bills were neatly folded inside a note that said, simply, "For the beautiful animals." Our tally for the day was $41,302; our tally for the first two weeks of the campaign was just over $450,000. Plenty of wonderful people did their part.

This is the foundation upon which we will begin Chapter Two: a strong team to manage our growing sanctuary with love and skill, and funds generously provided by supporters from around the country to sustain the components of our expanding program. I am grateful for, and humbled by, our success. Yet I'm also, I admit, a little lonely. The weight of what I've taken on often feels heaviest in wintertime. I'm grateful to have the yellow mutt, my best friend, by my side, thirteen and going strong. I bend to kiss him. "Come on, pup," I say. "Gotta go to work."

The first part of our expansion—to get ready for more animals— is straightforward: build more barns, create more pastures, hire additional staff to care for the animals. As the funding comes in, we get to work. We build a beautiful barn for pigs, two new barns for goats and sheep, and a sixty-foot "chicken chalet" for some lucky birds. Our carpenter, Caleb Fieser, is an artist; the chicken barn, his first CAS creation, is beautiful. I would happily live in it. The pigs are given a large square pasture close to the barn on low ground and bisected by a seasonal stream. In the middle, we dig a spacious wallow. David and the two Franks—Frank Sr., the former owner of our farm, and his son, Frank Jr., owner of Tiano Excavation, create two spacious pastures on high ground for new goats and sheep. How happy they are on their hilly terrain! How happy we are to provide deluxe digs for deserving animals.

The second part of our expansion—choosing which animals to accept—is excruciating. Historically, most of our animals have come from hoarders, and one of the most satisfying parts of my job has been to work with law enforcement to remove profoundly neglected animals from such people. Receiving them is wonderful, unambiguous joy, since our only role is to nurture wounded bodies and shattered spirits back to health.

But lately, the situation stacks up differently. Lately it is not the police or the SPCA knocking on our door, telling us about two hundred starving animals on an upstate hoarder's derelict farm. Rather, it's two hundred *individuals* knocking at the door because they can no longer provide for their beloved pets. One woman has suddenly lost her new husband to a heart attack; she has neither the expertise nor the means to support his horse. A second woman has lost her job and must surrender the horse she rescued from a yard sale, a "free" sign hung around its neck. A couple who adopted two horses from us several years earlier must bring them back. The husband is gravely ill; medical bills have wiped them out.

The stories leave us reeling: one job loss story after another, one foreclosure after another. A young couple, both of whom have lost their jobs, is moving to Tennessee to live with relatives. They appear in our driveway with Pinky and Peggy Sue, their beloved pet pigs, crammed into the back of a U-Haul with all their furniture. "Please, we've gone everywhere," the young wife pleads. "We can't take them with us." We are *packed* with pigs, but these girls are already close to hyperventilating from riding in a windowless van stuffed floor to ceiling with the detritus of a life. We say yes to the morbidly overweight, unspayed pigs, and an hour later, volunteer Paula Pell is lying in their freshly bedded stall singing softly to the newcomers. Meanwhile, Julie adds four more horses to the waiting list, and all around the region, sanctuaries are all either at or above capacity.

Yes, determining which animals to take in is always, *always* an excruciating choice. How is one farm foreclosure more dire than the next? One person desperately needs to find a home for her five sheep, four goats, three horses, and a rescued veal calf; a second needs to place four cows, four horses, a donkey, and a "coop full" of chickens (she thinks about thirty). Eeny, meeny, miney, mo.

Choosing Your Battles

In the last few years, America has turned a spotlight on agribusiness. Thanks to the success of such books as Matthew Sculley's *Dominion* and Jonathan Safran Foer's *Eating Animals,* to sustained campaigns by the big guns within the animal welfare and animal rights communities, and to such programs and films such as *Death on a Factory Farm* and *Food, Inc.,* it's becoming increasingly difficult for ordinary Americans to filter out the suffering animals endure as they are grown and killed for us to eat them. Millions have been introduced to the concept of downed animals and to the enormity of their suffering. Millions now know that commercially grown food animals (some 98 percent of the animals grown to feed us) know nothing but suffering and terror from birth to death. While we humans are still masters of denial and compartmentalization, it's becoming increasingly difficult to utter the words I've heard hundreds of times—"but we have laws to protect the animals!"—without sounding either stupid or something far more malevolent.

Even such icons as Oprah Winfrey and Martha Stewart have helped make the words "vegetarian" and "vegan" a little more comfortable for mainstream folks. Oprah tried a three-week vegan diet and blogged about the experience, and in 2009, Martha dedicated an entire episode of her show to a vegetarian Thanksgiving. Ellen DeGeneres does far more to advance the cause: she and her partner are vegans, and Ellen regularly features animal advocates on her show.

Gestation crates. Veal crates. Tens of thousands of chickens grown in a single building, sitting in their own filth, the ammonia from their excretions damaging eyes, lungs, and throats. Grown so quickly that the chicken industry now knows the percentage of birds that die of violent heart attacks because their organs can't handle the stress of being forced to grow to slaughter weight in just six weeks. People chop off animals' tails, chop off their toes, chop off their beaks, all without anesthesia, because on modern farms, animals packed together peck and claw and bite each other . . . and god forbid the meat be damaged. Every aspect of agribusiness is designed to maximize efficiency and profit despite the costs to the "product" (in this case, living animals) and the consumer. Ordinary Americans understand this. There's more work to be done, but the information is out there, readily and frequently available, in national media and hometown newspapers, on blogs that get 100,000 hits a day and little local radio stations.

Likewise, the impact of growing 65 billion animals annually to feed a human population of 6 billion is wreaking environmental havoc, and that fact, too, is becoming common knowledge. Earth is hot and getting hotter, and climatic devastation is here. Methane from cows; the razing of the ecologically important Amazon to graze cattle or to grow what feeds them; the melting of the North Pole due to warm air being trapped by greenhouse gases; the killing of our waterways by the waste of pigs, chickens, and cows; the extinction of species—plenty of us understand that growing animals to feed people is devastatingly inefficient and even more devastatingly harmful.

I think most of us even know that the consumption of animal products is killing us. Certainly the link between many forms of cancer, heart disease, stroke, obesity, type 2 diabetes, gout, high cholesterol, and other diet-related health problems Americans suffer from has been part of the public conversation, despite the fact that the medical establishment and the pharmaceutical industry might prefer it not be. They make money from our sickness. When we're well, we don't need them, and increasing numbers of us now understand what noted physicians and researchers such as John MacDougall, T. Colin Campbell, Neil Barnard, and so many others have been telling us for so long: the only healthy diet is a vegan diet.

But we're still eating animals—around ninety animals a year per American. Yes, of course I realize that changing one's diet is a tall order. But even though "quitting meat and dairy" might be every bit as difficult as quitting smoking for some people, it's at least as important, if not more so, since a smoker arguably injures only himself and those others inhaling second-hand smoke. As an organization that exists in part to raise awareness of agribusiness and encourage people to consider kinder diets, Catskill Animal Sanctuary struggles constantly with how best to help people really "get it."

A week after my first book was released in paperback, I was invited to give a reading at a local library. The audience numbered about sixty people, ranging in age from eight to eighty. I was delighted. It had been two and a half years since the first hardcover printing, and during that time I had formulated an approach to discussing what still remains a largely taboo subject: the consumption of animals. I hoped I'd have an opportunity to try it out tonight.

After introducing myself and offering a brief overview of Catskill Animal Sanctuary, I read the chapters that had become my standards. They're the chapters that disarm people. One depicts the night that Paulie, my rooster friend (who was spending a bitterly cold night in a crate inside my house because I'd found him in the barn shivering), crowed incessantly until, unable to sleep, I climbed from bed, lifted him out of the crate, and placed him on the pillow beside my head. No, I'm not making this up; he made it quite clear that this was *exactly* what he wanted. The next morning, I woke before he did. Paulie was snuggled deeply into his plush pillow. He hadn't moved an inch.

I read a chapter titled "One Cold Bitter Night," which describes the night that Rambo, a violent, dangerous sheep when we first took him in from a cruelty case, told me that we'd left two turkeys outside in the cold. On this particular night, Rambo had known our turkeys were out in the cold, had figured out a way to let me know, had understood that I would help. What stopped me in my tracks, though, was that he had cared about two animals of another species. In a single act that I will never forget, Rambo demonstrated a remarkable awareness of his surroundings, sophisticated problem solving, an ability to communicate a complex problem to another species, an awareness of both my role and the purpose of Catskill Animal Sanctuary, and *empathy.*

That single night changed forever my perception of the supposed differences between humans and other animals. It helped me understand—unequivocally—that they feel and know far more than most of us would ever believe. That they have far greater capacity than we humans, who rarely interact with and observe animals other than those who share our homes, realize. This chapter, in particular, is an important educational tool.

At the end of the reading, hands shot into the air. A woman asked how many animals we'd rescued, and how many currently live at Catskill Animal Sanctuary. A teenager asked if I had a favorite species, and if so, why that was the case. Someone else asked what inspired me to do this work. A fourth person asked whether pigs were as smart as she'd always heard. The questions continued. No one, though I always invite such a conversation in my opening talk, wanted more information about the link between agribusiness and global warming. Not one person asked about diet, though I'd made an extra effort to encourage such questions tonight. But finally, someone asked, "What's the most difficult part of your work?" and I had my opening.

"It's not what you think," came my response, followed by a challenge: "What would you guess it is?"

"Not hating people," a woman offered.

"Raising enough money to keep the sanctuary going," suggested another.

"Working with animals who are scared and dangerous," came an insightful comment.

"Actually, that's the greatest joy of what we do," I said. "I wish I had more time to do it like I did when we were a small organization." Heads nodded in understanding.

Finally, the guessing was over, and it was time.

"Talking to audiences like you is the hardest part of my job," I said.

Everyone laughed, and someone made the joke about public speaking being a bigger fear than death. Despite my own nervousness at entering uncharted territory, I inched forward.

"I'm going to invite you to look at a very popular topic from a unique and personal point of view," I said. "I hope you'll stay with me through this exercise."

"How many of you consider yourselves kind people?" I asked.

Nearly all hands went up.

"Raise your hand if you consider yourself a loving person."

Again, nearly everyone raised her hand.

"Raise your hand if you consider yourself someone who's strongly opposed to cruelty to animals."

Instantly, every single hand went into the air.

"Raise your hand if you believe pain feels the same to an animal as it does to a human." Again, every single hand went up, though I sensed some hesitation, and clearly some nervousness.

"Raise your hand if you're vegetarian."

Eight or nine hands went up, then I said to that group: "Fish are animals," and a teenager and several women lowered their hands. I wanted to say, "Milk and cheese come from animals," but I was concerned I'd lose the audience. Already five people were on their way out the door.

"Okay, kind and loving people who hate cruelty, I have a question for you: Why do you eat animals?" There. My heart was racing, my palms were sweating, but I had done it.

I'm convinced that meat eaters and vegetarians rarely have this conversation. While we humans love to talk about food, this particular dialogue just doesn't happen. Certainly as cultural mores begin to shift and there's broader recognition of the compelling reasons to adopt a diet free of animal products, people who eat animals—still the vast majority of us—are beginning to feel defensive of their food choices, at least around those of us who choose differently. On the other side, for reasons that I can't fathom, far too many vegans are judgmental and angry, choosing to overlook the fact that for most of their lives, they, too, ate meat and dairy. How dare we judge someone who hasn't yet begun the journey? How much more important it is for us to find inviting ways to engage others in a conversation that's become truly urgent, and to provide the skills and confidence to help them consider eating with conscience.

Tonight was my first stab at that effort. We talked until closing time, and our wonderful exchange touched on denial, excuses, questions of whether one could truly be healthy on a vegan diet, how and why I've made my own dietary choices over the years, and what it would take to convince audience members to change their diets. Among the most enlightening moments for me was the range of responses to the question "Why do you eat meat if you know the consequences of doing so?" People offered the following:

1. What do you mean by consequences? (I offered a quick refresher on the suffering of the animals and the planet.)

2. It's easier to slap together a ham and cheese sandwich than to make a salad—I'm lazy!

3. I don't have time to cook, and fast food restaurants are all about burgers (someone in the audience pointed out how one can purchase salads, veggie burgers, and baked potatoes at fast food stops).

4. I try not to think about this stuff: I like my meat!

5. Because you don't really think of what you're eating as an animal. "Meat" is meat. Somehow it doesn't feel like it was ever really a living animal.

6. I wouldn't know what else to cook.

7. It's hard to break a habit.

8. My husband would never eat "that other stuff." (I presume she meant "stuff" like tofu, tempeh, and seitan, things I almost never eat either because of the degree of processing involved in making them.)

9. This is not my cause. I love animals, but I believe they were put here for us humans to use for our own survival.

10. A range of responses that in essence meant "Eating meat is part of my identity. It's part of my family traditions, my cultural identity, my holiday celebrations, my family history. If I gave up _____ (fill in the blank with your five favorite meat recipes), *I wouldn't feel like myself: I'd feel like I'd lost something.* One woman commented that thinking about changing her diet almost felt like thinking about changing her children.

Yes, absolutely, this is the hardest part of my job—encouraging people to look animal consumption squarely in the face, to have the courage to address their role in the suffering, and to assess honestly whether they can still legitimately call themselves *kind* if they're knowingly contributing to such horrific suffering. Yes, it is hard as hell to have a "loving, supportive confrontation," of sorts, with groups of absolute strangers—but necessary. The stories of the rescue and healing of Noah, Andy, Norma Jean, and the like are the "feel good" part of this book. And while emergency rescue will *always* be at the heart of what we do, the fact is that if you choose to give up the consumption of animal products, *you alone* will save the lives of thousands of animals. You alone will spare enormous suffering of thousands of animals who, at the heart of it, are very little different from you and me.

If nothing else, I hope this book has convinced you of that.